KT-231-805
WITHDRAWN

WAVELETS
AND THEIR APPLICATIONS

CASE STUDIES

EDITED BY MEI
KOBAYASHI

IBM Tokyo Research Laboratory
Tokyo, Japan

Society for Industrial and Applied Mathematics

Philadelphia

10 9 8 7 6 5 4 3 2 1

Library of Congress Cataloging-in-Publication Data

Wavelets and their applications : case studies / edited by Mei Kobayashi.
p. cm.
Includes bibliographical references and index.
ISBN 0-89871-416-8 (pbk.)
1. Wavelets (Mathematics) I. Kobayashi, Mei.
QA403.3 .W362 1998
515' .2433--ddc21 98-3698

Contents

Preface

This book consists of a collection of essays on how wavelet techniques were used to solve open problems and to develop insight into the nature of the systems under study. The essays were contributed by my colleagues from a number of disciplines (mechanical and nuclear engineering; mathematical sciences; seismology; signal processing) whom I have had the privilege of meeting through conferences, tutorials, and seminars during the past few years of my stay in Japan.

Although calculus and some junior and senior mathematics courses for scientists and engineers will suffice, a solid background in undergraduate mathematics—particularly analysis and numerical analysis—and some familiarity with the basics of wavelets are helpful for reading this book. Readers interested in introductory and reference material on wavelets can consult a number of good texts. *Ondelettes et Algorithmes Concurrents*, *Ondelettes et Operateurs*, and *Operateurs de Calderon-Zygmund*, a three-volume set by Meyer and *Ten Lectures on Wavelets* by Daubechies are recognized as the mathematical authorities on the subject. Meyer's book assumes that readers have a strong background in pure mathematics. Daubechies' book is easier reading for applied mathematicians who have not had rigorous training in abstract operator theory, but it does require a strong background in real and complex analysis. Strang and Nguyen's *Wavelets and Filter Banks* is a more elementary introduction to the subject and is more appropriate for undergraduate math majors and engineers. A more advanced and specialized text from an engineering/signal processing perspective is Vetterli and Kovačević's *Wavelets and Subband Coding.* Chui's books, *Wavelets: A Tutorial in Theory and Applications* and *Wavelets: A Mathematical Tool for Signal Analysis*, Sakakibara's *Wavelets: A Beginner's Guide*, and Ashino and Yamamoto's *Wavelet Analysis* have been well received as good tutorial-style books in Japanese. Two nice, short, introductory essays on the basics of wavelets and implementation are Chapters 1 and 2 by Aldroubi and Unser in their coedited book, *Wavelets in Medicine and Biology.* Good introductory essays in Japanese are given in the December 1992 issue of *Mathematical Sciences.* Sato's essay [48] (also in Japanese) in the

Journal of the Acoustical Society of Japan may be more suitable for engineers.

This book is written as a collection of essays in a case studies format. Each essay begins with a description of the problem under study and points to specific properties of wavelets and techniques which were used to determine a solution. In each case study, the goal is to quickly determine a simple solution to a specific problem at hand. Scientists do not necessarily need to develop a new theory or conduct extensive (and probably expensive) comparison studies—neither of which is an encouraged practice in industrial laboratories.

The first case study describes work conducted at the Earthquake Research Institute (ERI) of the University of Tokyo to develop a system to accurately display two-dimensional geographical data at user-specified resolutions on a personal computer (PC). Wavelet-based methods for curve and three-dimensional map display have been investigated by other scientists in different contexts. In particular, we mention Certain et al.; Eck and Hoppe; Finkelstein and Salesin; Gross et al.; Lindstrom et al.; Luebke and Erikson; Reissel; Stollnitz, DeRose, and Salesin; and Zorin, Schröder, and Sweldens, who developed very sophisticated graphics algorithms for displaying information. As beautiful as the graphics are, the computationally intense methods are too expensive for the simple needs of the scientists at ERI.

In the second case study, a very simple and inexpensive wavelet-based technique to reduce noise in data is successfully applied to correct experimental measurements of dry friction and data from a drop mass test. The attractiveness of the method lies in its relative simplicity and ability to run on a small, unremarkable PC; its application to mechanical engineering laboratory data yields results which are so clean that more involved and sophisticated methods, such as those by Coifman and Donoho; Hilton et al.; Johnstone; Malfait; Saito; and those by Sasaki and Yamada (described in Chapter 5) are not needed for the specific applications which were considered.

The third case study presents a powerful and simple new wavelet-based preconditioning method for solving large systems of linear equations. The preconditioner leads to accurate results while substantially reducing computation time and cost in simulations of fluid flow modeled by Poisson equations. Preconditioning using wavelet bases leads to improved performance, since conjugate gradient methods cannot prevent the exponential increase in computing time when the number of gridpoints is increased. The wavelet method developed in the case study builds on earlier, more primitive results by Beylkin, Glowinski, and Jaffard, which can only be applied to problems with highly restrictive conditions, e.g., periodic boundary conditions.

In the fourth case study, wavelet analysis is used in the development of a Japanese text-to-speech (TTS) system for PCs. During the past decade, the focus on TTS systems has shifted from large mainframe computers to very small PCs, including notebook PCs. Since PC users consider TTS to be a "cute" option which would be nice (but not critical) to have, product development teams must create TTS systems which occupy little memory and output relatively natural, comfortable sounding synthesized speech. The wavelet technologies which were developed during the course of the case study helped to reduce memory requirements for the system's speech synthesis dictionary and to substantially improve the quality of the synthesized speech. The TTS research team received the 1996 Technology Development Award of the Japan Acoustical Society for their work. The techniques developed in the study are extensions of those by Kadambe and Boudreaux-Bartels (pitch marking) and Hamon, Moulines, and Charpentier (overlap-add). Other notable wavelet-based pitch-marking techniques have recently been developed by Kawahara and Cheveigine, and Yip, Leung, and Wong.

The fifth case study presents new wavelet techniques developed for and applied to the study of atmospheric wind, turbulent fluid, and seismic acceleration data. In one study, scientists at the Disaster Prevention Research Institute of Kyoto University analyzed how the topography of the region surrounding a large, man-made structure in Japan influences turbulent and potentially violent characteristics of high winds. More specifically, they developed a method for classifying wavelet transform coefficients from two distinct components of turbulent wind data. In a second study, wavelet techniques were used to study Navier–Stokes equations. The scientists found that the probability density function (PDF) of the wavelet coefficients of the velocity data has a power-form in regions with very high wavenumbers; more precisely, in the deep dissipation range. In a third study, scientists at the University of Tokyo and Kajima Corporation (one of the leading construction companies in Japan) developed a new method to clean noisy seismic acceleration data. Wavelet-based time-frequency analysis and expansions were used to identify and separate what were inextricably intertwined signal components and enabled correction of noisy data, which could not be easily cleaned using conventional methods.

The case studies in this book can be read independently of one another. They are part of a collection of works which were featured in the *Wavelets in Japan* series published by *SIAM News* [17]. The series highlighted the applicability of wavelet-based technologies to a variety of disciplines and underscored the importance of identifying suitable application scenarios for new mathematical tools.

Some of the more unusual scenarios which were also highlighted in *SIAM News* include application of a wavelet-based grid generator at Mitsubishi Heavy Industries by Jameson [12]; identification of irregularities in cement mixes by Aizawa et al. [1] at Chichibu Onoda, a leading Japanese cement company; identification of irregularities in cooling system valves, the development of a wavelet-based time-frequency analyzer with a user-friendly Japanese interface, and the development of a system to compress *keisoku* (control and sensor) data in a manner that preserves features which are essential for automatic, pattern-matching based identification of irregularities in data by Kazato et al. at Yamatake–Honeywell [13], [14]; identification of problems associated with assembly line conveyer belts (roller slippage and object drift) by Kitagawa and his colleagues at the Toyohashi University of Technology and a Japanese manufacturer [10], [16]; identification and diagnosis of automobile engine defects by Kikuchi and Nakashizuka of Niigata University and a leading Japanese automobile manufacturer [15]; and the development of golf balls which emit a more professional and "sporty" sound when they are struck and driven off a tee at angles that have been targeted by pros, by Zhang et al. at the Industrial Technology Center of Okayama Prefecture and a Japanese golf equipment manufacturer [26], [27].

Other books written from an applications perspective that I highly recommend are *Wavelets for Computer Graphics*, by Stollnitz, DeRose, and Salesin, and *Wavelets in Medicine and Biology*, edited by Aldroubi and Unser. Both are easy to follow and inspire the reader's imagination. *Wavelets and Their Applications*, edited by Ruskai et al.; *Wavelets: Mathematics and Applications*, edited by Benedetto and Frazier; *Wavelets: Theory and Applications*, edited by Erlebacher, Hussaini, and Jameson; and *Multiscale Wavelet Methods for Partial Differential Equations*, edited by Dahmen, Kurdila, and Oswald are nice collections of papers, both tutorial and advanced.

So many other excellent applied works on wavelets have been presented in conference proceedings, journals, and dissertations that I find it impossible to list them all. Some good starting points are the special issues of *IEEE Transactions on Information Theory* (March 1992); *Mathematical Sciences* (in Japanese, Dec. 1992); *IEEE Transactions on Signal Processing* (Dec. 1993); *Proceedings of the IEEE* (April 1996); *IEICE Transactions* (in Japanese, Dec. 1996); and recent special issues on wavelets in the *Journal of the Japan Society for Simulation Technology* (Dec. 1997); *IEEE Transactions on Circuits and Systems* (to appear, 1998); *IEEE Transactions on Signal Processing* (Jan. 1998); and proceedings from ACM Siggraph, AMS, IEEE ICAASP, and ICIP, IEICE, SIAM, and SPIE conferences. The *Wavelet Digest* at http://www.wavelet.org, a newsgroup dedicated to the subject, may be of interest to Internet buffs. I leave it to readers to hunt out these treasures and more.

It is my hope that this book will inspire students and researchers to develop and incorporate wavelet technologies into their repertoire when relevant and useful. At the same time, I would like to caution readers not to expect or claim that wavelets will always yield the best solution or lead to miracles. The exceptionally promising and coveted status that wavelets enjoyed in the late 1980s to early 1990s faded for good reason. Media distortion and exaggerated claims by some irresponsible and dogmatic proponents were misleading. Fortunately, the passing of time leads us to a more balanced perspective of the field; some good, practical applications have been developed in areas not anticipated, while other seemingly promising areas have borne little fruit. Although not as dynamic as the early days of the *"Ueburetto Boom"* in Japan and abroad, the contributions of wavelet technologies are a small but solid new subset of a far greater collection of classical and emerging technologies that are enhancing the quality of life for all.

In closing, I would like to express my gratitude to some of my colleagues and friends without whom this book could not have been possible. They are Ole Hald, my dissertation advisor; Sakae Uno and Kazutoshi Sugimoto (for their thoughtful management and support of my work on wavelets at the IBM Tokyo Research Laboratory); Michio Yamada (for his friendship and patient mentorship); Hirotsugu Inoue and Susumu Sakakibara (for organizing and inviting me to participate in the wavelet seminar series sponsored by the Japan Society of Mechanical Engineers); Hisakazu Kikuchi, Takahiro Saito, and the members of the IEICE wavelet study group (for inviting me to participate in their activities); Gail Corbett, Rena Bloom, and Carol Mehne (for their thoughtful and patient editing of my essays for *SIAM News*); and Cormac Herley, Yasuhiko Ikebe, and Dong-Shen Cai (for their encouragement and hospitality). I would also like to extend my gratitude to the contributors of this book for their amiable, patient cooperation and help and for allowing me great freedom in editing their work. Finally, I would like to thank my mother, Yukiko Kobayashi, who has always seemed to have more faith in me than I have in myself.

Mei Kobayashi, Tokyo, Japan September 1997

References

[1] T. Aizawa, N. Takechi, T. Kozuki (1995), "Control and monitor of cement production," pp. 40–41 in *Electronics Magazine* (in Japanese).

[2] A. Aldroubi, M. Unser (eds.) (1996), *Wavelets in Medicine and Biology*, CRC Press, New York.

[3] R. Ashino, S. Yamamoto (1997), *Wavelet Analysis*, Kyoritsu, Tokyo (in Japanese).

[4] J. Benedetto, M. Frazier (eds.) (1993), *Wavelets: Mathematics and Applications*, CRC Press, New York. (Also available: Japanese translation by M. Yamaguti and M. Yamada, of selected works of Benedetto and Frazier, Springer-Verlag, Tokyo.)

[5] C. Chui (1992), *An Introduction to Wavelets*, Academic Press, Tokyo. (Also available: Japanese translation by A. Sakurai and T. Sakurai, Tokyo Denki Daigaku, Tokyo.)

[6] C. Chui (1997), *Wavelets: A Mathematical Tool for Signal Analysis*, SIAM, Philadelphia, PA.

[7] W. Dahmen, A. Kurdila, P. Oswald (eds.) (1997), *Multiscale Wavelet Methods for Partial Differential Equations*, Academic Press, Tokyo.

[8] I. Daubechies (1992), *Ten Lectures on Wavelets*, SIAM, Philadelphia, PA.

[9] *Electronics Magazine* (1995), Nov., Special issue on wavelets, Ohmusha Ltd. (in Japanese).

[10] H. Horihata et al. (1996), "Application of fractal analysis for surface flaw inspection of steel sheet," pp. 26–34 in *Nondestructive Characterization of Materials IV*, R. Green, K. Kozaczek, C. Runn (eds.), Trans Tech Pub., Zurich, Switzerland.

[11] G. Erlebacher, M. Hussaini, L. Jameson (eds.) (1996), *Wavelets: Theory and Applications*, Oxford Univ. Press, Oxford, UK.

[12] L. Jameson, "A wavelet-optimized, very high order, adaptive grid and order numerical method," *SIAM J. Sci. Comput.*, 19 (1998), pp. 1976–2009.

[13] H. Kazato, Y. Hiraide (1997), "Estimation of dead time using correlation analysis in the time-frequency domain," *Proc. Sympos. System Identification (SYSID'97)—July 8–11, Fukuoko, Japan.* Session: Identification through Wavelet. Intl. Federation of Automat. Control, pp. 27–37 (http://www.ifac-control.org/index.html//).

[14] H. Kazato, T. Hosoi (1995), "Data compression for measured variable by wavelet transform," *Proc. 10th Digital Signal Processing Symp.*, IEICE, Nov. 1, pp. 205–210 (in Japanese).

[15] H. Kikuchi et al. (1992), "Fast wavelet transform and its application to detecting detonation," *IEICE Trans. Fund.*, E-75-A, pp. 980–987.

[16] H. Kitagawa (1997), "Wavelet and fractal analysis of assembly line data," (lecture), Wavelet Seminars, sponsored by the Japan Society of Mechanical Engineering, Aoyama Gaku-in, Tokyo, March 28 (in Japanese).

[17] M. Kobayashi et al., "The Ueburetto Boom" series, *SIAM News*: "The ueburetto boom," pp. 18 & 24, Nov. 1995; "Listening for defects," pp. 18 & 24, March 1996; "Wavelets for a modern muse," pp. 8–9, May 1996; "Netting Namazu," pp. 15 & 20, July/Aug. 1996; "Wavelets and the gorufu boom," pp. 4 & 11, July/Aug. 1997.

[18] Y. Meyer (1990), *Ondelettes et Algorithmes Concurrents*, *Ondelettes et Operateurs*, and *Operateurs de Calderon–Zygmund*, vols. 1–3, Hermann, Paris.

[19] M. Nakashizuka (1996), "Wavelet analysis of medical data," (lecture), Workshop on Wavelets and Their Applications, IBM Tokyo Research Laboratory, July.

[20] M. Ruskai et al. (eds.) (1992), *Wavelets and Their Applications*, Jones and Bartlett, Boston, MA.

[21] S. Sakakibara (1995), *Wavelets: A Beginner's Guide*, Tokyo Denki Daigaku, Tokyo (in Japanese).

[22] M. Sato (1991), "Mathematical foundation of wavelets I and II," *J. Acoust. Soc. Japan*, vol. 47, pp. 405–423 (in Japanese).

[23] E. Stollnitz, T. DeRose, D. Salesin (1996), *Wavelets for Computer Graphics*, Morgan-Kaufmann, San Francisco, CA.

[24] G. Strang, T. Nguyen (1996), *Wavelets and Filter Banks*, Wellesley–Cambridge Press, Wellesley, MA.

[25] M. Vetterli, J. Kovačević (1995), *Wavelets and Subband Coding*, Prentice–Hall, Englewood Cliffs, NJ.

[26] Z. Zhang (1997), "Wavelet analysis and golf ball manufacturing," (lecture), Wavelet Seminars, sponsored by the Japan Society of Mechanical Engineering, Aoyama Gaku-in, Tokyo, March 28.

[27] Z. Zhang, H. Kawabata (1996), "Signal analysis by using spline wavelet," *Proc. Seigyo Joho Gakkai*, 2nd ASP Symposium, pp. 21–24 (in Japanese).

1. Wavelet-Based Multiresolution Display of Coastline Data

Sumiko Hiyama* and Mei Kobayashi†

Abstract. This chapter presents results from a case study on developing a system to accurately display two-dimensional geographical data on a workstation or personal computer (PC) for use by the scientists at the Earthquake Research Institute (ERI) of the University of Tokyo. Scaling function and wavelet expansions are used to display Japanese coastline data at various resolution levels. During the development of the system, we assessed the ability of four types of wavelets to model curves and found that Coiflets and Daubechies' symmetric, biorthogonal wavelets of degree two are particularly effective for modeling intricate as well as broad features of coastlines. The technique we use for coastline display can be extended to more general use in curve approximation and design on workstations and PCs.

Key words. wavelets, multiresolution analysis, multiresolution display, approximation, curve, coastline

1.1. Introduction

This chapter presents results from a case study on developing a system to accurately display two-dimensional geographical data on a personal computer (PC) for use by the scientists at the Earthquake Research Institute (ERI) of the University of Tokyo. Some preliminary results from this work appear in [16]–[18]. The problem can be stated more precisely as follows: *Given Japanese coastline point data, display a view of the coastline at various resolution levels as quickly and accurately as possible on a workstation or PC.* We model the coastline as a curve in two-dimensional space and reduce the problem to one of curve approximation and display. The amount

*Earthquake Research Institute, The University of Tokyo, Yayoi 1-1-1 Bunkyo-ku Tokyo 113-0032 Japan (hiyama@eri.u-tokyo.ac.jp).

†IBM Tokyo Research Laboratory, 1623-14 Shimotsuruma, Yamato-shi, Kanagawa-ken 242-8502 Japan (mei@trl.ibm.co.jp).

of available laboratory data is too large to be shown in full detail for the resolution of a PC screen, so a method for quickly extracting useful, reduced amounts of data and generating an approximation for a curve is needed.

In computer graphics, a curve is usually described as a sequence of connected line segments, with points of intersection $\{p_0, p_1, \ldots, p_n\}$. There are two ways to reduce the amount of data for describing a curve: a subset of the points $\{p_0, p_1, \ldots, p_n\}$ can be used to obtain a coarser approximation of a curve [9], [21], [26], or a linear combination of a suitable set of functions can be used (see Figure 1.1). We take the latter approach and use scaling functions and wavelets as basis functions to solve the coastline display problem.

Traditionally, fractal curves [1], [2], [25] and splines [3], [4], [11], [29] have been used to approximate curves; however, these sets do not readily lend themselves to the resolution of our problem. Either fractal dimensions of intricate curves representing coastlines must be calculated or optimal placement of knots for splines must be determined, both of which require extensive and tedious computations. Wavelet methods are well suited for approximating coastlines for several reasons: scaling functions and wavelets are localized in space so that sharp spikes can be represented easily and accurately; the fractal-like features of some types of scaling functions and wavelets resemble intricate zig-zags found in coastlines; scaling functions and wavelets can be used to approximate broad cusps in coastlines by setting the appropriate coefficients in the expansion to be very large; and computations to move through various resolution levels using dyadic wavelets can be made very efficient by using techniques analogous to fast Fourier methods. We studied four types of scaling functions and wavelets that have the features described above to determine their effectiveness in approximating coastlines.

Some other works, with similar approaches to ours, that use wavelets for curve and coastline approximation and display are [12], [27], and [30]. Visualization studies to display three-dimensional representations of data, interactively and at various resolution levels using wavelets, are reported in [5], [10], [13], [19], [20], [22], [23], [37], and [38]; however, the techniques developed and used by the scientists require powerful computers with large memories, i.e., implementation on current, standard-level PCs is not a realistic possibility.

This chapter is organized as follows. The next section is a review of basic concepts and properties associated with wavelets which are relevant for solving the problem proposed in our case study. Then, we outline an algorithm proposed by Mallat (for image processing applications) which uses scaling functions and wavelets to approximate curves at vari-

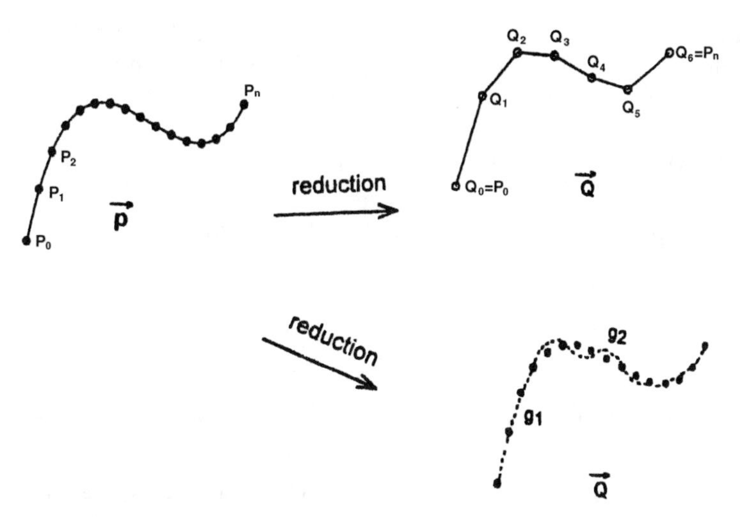

Figure 1.1. *Two approaches to reducing data.*

ous resolution levels. Next, we discuss implementation experiments using four types of wavelets with compact support: Daubechies' orthonormal, Coiflets, Daubechies' biorthogonal, and biorthogonal splines. Properties of the wavelets are given, along with results from approximating a spiral curve, a kuri curve, and portions of the Japanese coastline. Finally, conclusions from our experiments are discussed, directions for future work are outlined, and possible extensions of the technique to other applications are given.

1.2. Properties of Scaling Functions and Wavelets

In this section we describe properties of scaling functions and wavelets and how they can be exploited for approximating curves. The concept of a multiresolution analysis and its relation to expansions of functions using scaling functions and wavelets as bases is a central tool for solving our problem. The definition given below and related discussions follow from those in [7] and [24].

Definition. Closed subspace families $\{V_j : j \in \boldsymbol{Z}\}$ of $L^2(\boldsymbol{R})$ are said to constitute a *multiresolution analysis (MRA)* when the following five conditions are satisfied:

(i) $\ldots \subset V_2 \subset V_1 \subset V_0 \subset V_{-1} \subset V_{-2} \subset \ldots,$

(ii) $\overline{\bigcup_{j\in\boldsymbol{Z}} V_j} = L^2(\boldsymbol{R}) \quad \text{and} \quad \bigcap_{j\in\boldsymbol{Z}} V_j = \{0\},$

(iii) $\varphi(x) \in V_j \iff \varphi(2x) \in V_{j+1}; \qquad \forall j \in \boldsymbol{Z},$

(iv) $\varphi(x) \in V_j \implies \varphi(x - 2^{-j}k) \in V_{j+1}; \qquad \forall k \in \boldsymbol{Z},$

(v) $\{\varphi(x-k) : k \in \boldsymbol{Z}\}$ form an orthonormal basis for V_0.

$\varphi(x)$ is called the scaling function of the .

Given any , there always exists an orthonormal wavelet basis $\{\psi_{j,k} : j, k \in \boldsymbol{Z}\}$ of $L^2(\boldsymbol{R})$ such that $\psi_{j,k}(x) = 2^{-j/2}\psi(2^{-j}x - k)$, and

$$P_{j-1}f = P_j f + \sum_{k\in\boldsymbol{Z}} \langle f, \psi_{j,k}\rangle\, \psi_{j,k}, \quad \forall\, f \in L^2(\boldsymbol{R}), \tag{1.1}$$

where f is a function in $L^2(\boldsymbol{R})$ and P_j is the orthogonal projection onto V_j. For every $j \in \boldsymbol{Z}$, let W_j be the orthogonal complement of V_j in V_{j-1}, then $V_{j-1} = V_j \oplus W_j$, and $W_j \perp W_{j'}$ for $j \neq j'$ so that

$$V_j = V_J \oplus \bigoplus_{k=0}^{J-j-1} W_{J-k} \quad \text{for} \quad j < J, \tag{1.2}$$

and

$$L^2(\boldsymbol{R}) \;=\; \bigoplus_{j\in\boldsymbol{Z}} W_j. \tag{1.3}$$

In other words, for fixed j, the set $\{2^{j/2}\varphi(2^j x - k) : k \in \boldsymbol{Z}\}$ forms an orthonormal basis for V_j and $\{2^{j/2}\psi(2^j x - k) : k \in \boldsymbol{Z}\}$ forms an orthonormal basis for W_j. As a consequence, any function f in $L^2(\boldsymbol{R})$ can be approximated by a linear combination of the scaling functions $\{2^{j/2}\varphi(2^j x - k) : k \in \boldsymbol{Z}\}$ in V_j. If we move from V_j to a coarser resolution level V_J, then we can expand f using the functions $\{2^{J/2}\varphi(2^J x - k) : k \in \boldsymbol{Z}\}$. The difference in the approximations of f at two successive resolution levels can be expressed as a linear combination of the wavelets, as shown in equation (1.1). We can determine the difference in the approximations of f at levels j and J by applying the formula $(J-j)$ times, in succession. If, instead, we move from a coarse resolution level to an infinitesimally fine resolution level, we will be able to approximate all functions in $L^2(\boldsymbol{R})$, as indicated by equations (1.2)–(1.3).

The concepts discussed above are illustrated through the following example. Let $\psi(x) \in W_0$ and $\varphi(x) \in V_0$. These functions can be expressed as a sum of basis functions $\{\varphi(2x-k) : k \in \boldsymbol{Z}\}$ of V_1 as

$$\varphi(x) = \sum_k \alpha_k \sqrt{2}\varphi(2x-k) \quad \text{and} \quad \psi(x) = \sum_k \beta_k \sqrt{2}\varphi(2x-k), \tag{1.4}$$

where $\beta_k = (-1)^k \alpha_{1-k}$. The α_k and β_k are called *filter coefficients*, and they characterize the different wavelet families. (Some recommended tutorials and references on filter banks are found in [15], [28], [31], [33], [34], [35], and [36].) Equation (1.4) tells us that if V_1 is a closed subspace, then any $f(x) \in V_1$ can be expanded to a sum of scaling functions $\varphi(x)$ of V_0 and wavelet functions $\psi(x)$ of W_0 as

$$f(x) = \sum_l C_l^0 \; \varphi(x-l) + \sum_l D_l^0 \; \psi(x-l), \tag{1.5}$$

where

$$\begin{aligned} C_l^0 &= \int f(x)\; \varphi(x-l)\; dx \\ &= \int \sum_m C_m^1 \; \sqrt{2}\; \varphi(2x-m) \sum_k \alpha_k \sqrt{2}\; \varphi(2x-2l-k)\; dx \\ &= \sum_m \alpha_{m-2l} \; C_m^1 \;, \end{aligned}$$

and

$$D_l^0 \;\; = \;\; \sum_m \beta_{m-2l} \; C_m^1,$$

since $V_1 = V_0 \oplus W_0$. The coefficient $\mathbf{C}^1$ can be expressed in terms of the $\mathbf{D}^0$ and $\mathbf{C}^0$ as

$$C_m^1 = \sum_l \alpha_{m-2l} \; C_l^0 + \sum_l \beta_{m-2l} \; D_l^0. \tag{1.6}$$

The algorithm outlined through this example was developed by Mallat, who applied it to the processing of digital images [24]. For each component, the

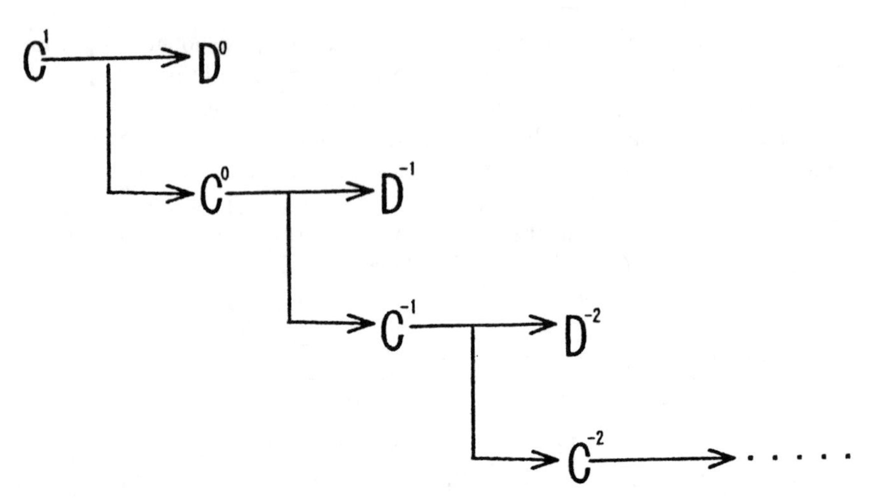

Figure 1.2. *Pyramidal hierarchy of C^i and D^i.*

operation count for one step of Mallat's algorithm is $(Nl/2)$, where N is the length of $\mathbf{C}^1$ and l the size of the filter $\{\alpha_k\}$. The $\mathbf{C}^i$ are known as *the level i scaling coefficients.* Equations (1.4)–(1.6) show us how we can approximate a function and move it through different resolution levels. The pyramidal hierarchy of $\mathbf{C}^i$ and $\mathbf{D}^i$ is illustrated in Figure 1.2. In the next section, we will take the function to be a curve and determine the effectiveness of the algorithm for curve approximation and display.

1.3. Approximation of Curves Using Mallat's Algorithm

In this section we show how to apply Mallat's algorithm to the approximation and display of a curve. First we describe four types of wavelets used in our experiments. We decided to use Daubechies' orthonormal wavelets and Coiflets in our experiments because they have the fractal-like features found in coastlines, and computations with them are simple and fast. Since these wavelets are not symmetric, computations starting at one end of a curve might yield very different results than those which start at another end. We decided to examine symmetric, biorthogonal wavelets to assess the differences in the results. All four wavelets are defined by an algorithm to compute them numerically from their filter coefficients. We use orthonormal

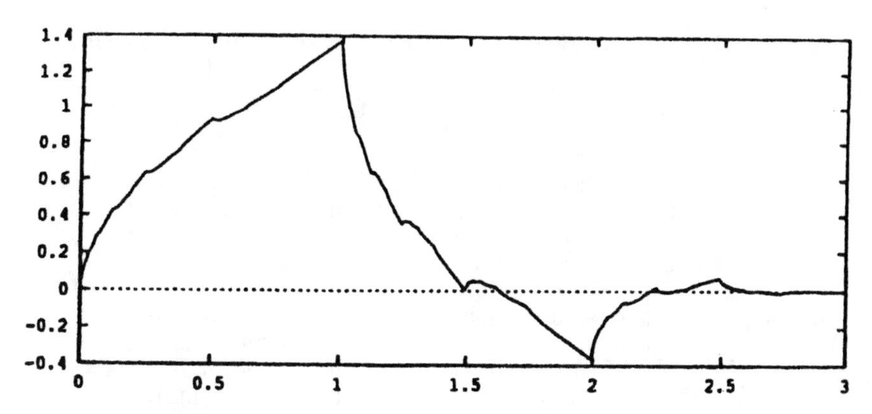

Figure 1.3. *Daubechies' scaling function for* $N = 2$.

wavelets with compact support to show how to implement the algorithm. The others follow similarly.

1.3.1. Daubechies' Orthonormal Wavelets

There is no closed formula for the orthonormal wavelets with compact support proposed by Daubechies [6]. These wavelets are defined by an iterative algorithm for computing function values dyadically using filter coefficients. The scaling functions are computed first in the algorithm. For any user-specified resolution, a wavelet is computed using the corresponding scaling function. The algorithm proceeds as follows.

Step 1. The scaling function $\varphi(x)$ is defined by

$$\varphi(x) = \lim_{l \to \infty} \eta_l(x),$$

where η_l is determined iteratively by

$$\begin{aligned} \eta_0(x) &= \chi_{[-1/2,1/2]}, \\ \eta_l(x) &= \sqrt{2} \sum_n \alpha(n)\, \eta_{l-1}(2x - n), \end{aligned}$$

where χ is the characteristic function, and $\alpha(i)$ are filter coefficients.

Table 1.1. *Filter coefficients for Daubechies' orthonormal wavelets.*

N	n	α	n	α
2	0	0.4829629131445341	1	0.8365163037378079
	2	0.2241438680420134	3	-0.1294095225512604
3	0	0.3326705529500826	1	0.8068915093110926
	2	0.4598775021184916	3	-0.1350110200102546
	4	-0.0854412738820267	5	0.0352262918857095
4	0	0.2303778133088965	1	0.7148465705529156
	2	0.6308807679298589	3	-0.0279837694168599
	4	-0.1870348117190931	5	0.0308413818355608
	6	0.0328830116668852	7	-0.0105974017850690
5	0	0.1601023979741929	1	0.6038292697971897
	2	0.7243085284377729	3	0.1384281459013203
	4	-0.2422948870663823	5	-0.322448695846387
	6	0.0775714938400457	7	-0.0062414902127983
	8	-0.0125807519990820	9	0.0033357252854738
6	0	0.1115407433501095	1	0.4946238903984531
	2	0.7511339080210954	3	0.3152503517091976
	4	-0.2262646939654398	5	-0.1297668675672619
	6	0.0975016055873230	7	0.0275228655303057
	8	-0.0315820393174860	9	0.0005538422011615
	10	0.0047772575109455	11	-0.0010773010853085
7	0	0.0778520540850091	1	0.3965393194818917
	2	0.7291320908462351	3	0.4697822874051931
	4	-0.1439060039285649	5	-0.2240361849938749
	6	0.0713092192668303	7	0.0806126091510831
	8	-0.0380299369350144	9	-0.0165745416306669
	10	0.0125509985560998	11	0.0004295779729214
	12	-0.0018016407040475	13	0.0003537137999745

Step 2. The wavelet function $\psi(x)$ with N vanishing moments is computed from the associated scaling function $\varphi_N(x)$ using the formula

$$\psi(x) = \sum_{n=0}^{2N-1} (-1)^n \, \alpha_N(-n+1) \, \varphi_N(2x-n).$$

Filter coefficients α for Daubechies' orthonormal wavelets are listed in Table 1.1. The list consists of values first computed by Daubechies and subsequently corrected by Hatano [14]. We note that the wavelets have N vanishing moments and a relatively small support of width $2N-1$; in other words, α_n is nonzero only for $n = 1, 2, \ldots, 2N-1$. Note that both are not symmetric (see Figure 1.3 of Daubechies' scaling function for $N = 2$).

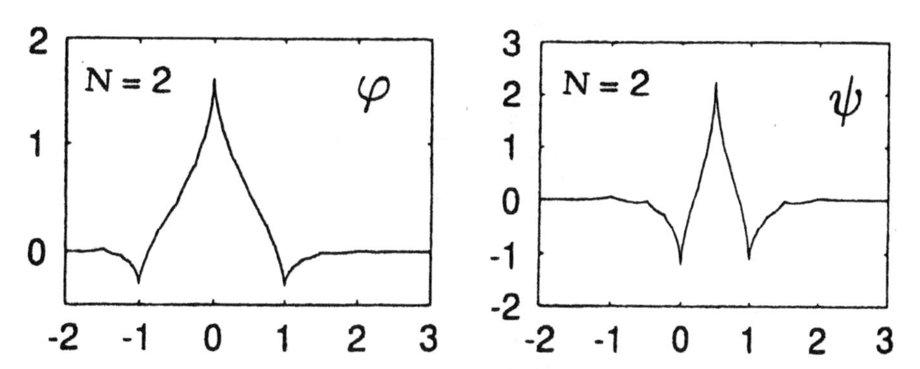

Figure 1.4. *Coiflets: Scaling functions and wavelets for* $N = 2$.

1.3.2. Coiflets

In the spring of 1989, R. Coifmann suggested the construction of orthonormal wavelets and scaling functions, both with vanishing moments, so when these wavelets were constructed, they were named "Coiflets" by I. Daubechies [8]. These wavelets must also be generated numerically from their filter coefficients since there is no closed formula for them. Coiflets appear more symmetric than Daubechies' orthonormal scaling functions and wavelets (see Figures 1.3 and 1.4), have compact support, but are not orthonormal. Furthermore, Coiflets ψ with N vanishing moments have associated scaling functions φ also with N vanishing moments, i.e.,

$$\int x^l \ \psi(x) \ dx = 0, \qquad l = 0, \ldots, N-1, \quad (1.7)$$

$$\int \varphi(x) \ dx = 1 \quad \text{and} \quad \int x^l \ \varphi(x) \ dx = 0, \qquad l = 1, \ldots, N-1. \quad (1.8)$$

Orthonormal wavelets satisfy condition (1.7), but the associated scaling functions do not necessarily satisfy (1.8). Coiflets do, however, have a major drawback: those with N vanishing moments have a much wider support of $6N - 1$, in contrast to $2N - 1$ for orthonormal wavelets. Filter coefficients α_k for Coiflets are listed in Table 1.2. The table consists of values first computed by Daubechies and subsequently corrected by Hatano [14].

1.3.3. Symmetric Biorthogonal Wavelets

As mentioned above, the two types of wavelets introduced thus far are not perfectly symmetric. This means that approximations for functions in which computations begin from the right-hand side will yield different results than those which begin from the left. To define a symmetric wavelet,

Table 1.2. *Filter coefficients for Coiflets.*

N	n	$\alpha/\sqrt{2}$	n	$\alpha/\sqrt{2}$
1	-2	-0.051429728471	-1	0.238929728471
	0	0.602859456942	1	0.272140543058
	2	-0.051429728471	3	-0.011070271529
2	-4	0.011587596739	-3	-0.029320137983
	-2	-0.047639590310	-1	0.273021046535
	0	0.574682393857	1	0.294867193696
	2	-0.054085607092	3	-0.042026480461
	4	0.016744410163	5	0.003967883613
	6	-0.001289203356	7	-0.000509505399
3	-6	-0.002682418671	-5	0.005503126708
	-4	0.016583560479	-3	-0.046507764479
	-2	-0.043220763560	-1	0.286503335274
	0	0.561285256870	1	0.302983571773
	2	-0.050770140755	3	-0.058196250762
	4	0.024434094321	5	0.011229240962
	6	-0.006369601011	7	-0.001820458916
	8	0.000790205101	9	0.000329665174
	10	-0.000050192775	11	-0.000024465734

the pair of filter coefficients α_k and β_k used to move to a coarser resolution level will be different from the pair $\tilde{\alpha_k}$ and $\tilde{\beta}_k$ used to move to a finer resolution level. The equations for biorthogonal wavelets corresponding to (1.4)–(1.6) from Mallat's algorithm are

$$C_l^0 = \sum_m \alpha_{m-2l}\, C_m^1,$$

$$D_l^0 = \sum_m \beta_{m-2l}\, C_m^1,$$

and

$$C_m^1 = \sum_l \tilde{\alpha}_{m-2l}\, C_l^0 \;+\; \sum_l \tilde{\beta}_{m-2l}\, D_l^0,$$

where $\beta_k = (-1)^k\, \tilde{\alpha}_{1-k}$ and $\tilde{\beta}_k = (-1)^k\, \alpha_{1-k}$.

We considered two types of symmetric biorthogonal wavelets in our experiments: splines and Daubechies' biorthogonal wavelets. Filter coefficients for biorthogonal spline scaling functions and wavelets were computed using formulae from Table 8.2 in [7]. The spline wavelets $\psi_{1,3}$ and $\tilde{\psi}_{1,3}$ were first constructed by Tchamitchian [32]. We used wavelets and scaling functions of degree $\tilde{N} = 3$ and $N = 1, 3, 5, 7, 9$ in our experiments.

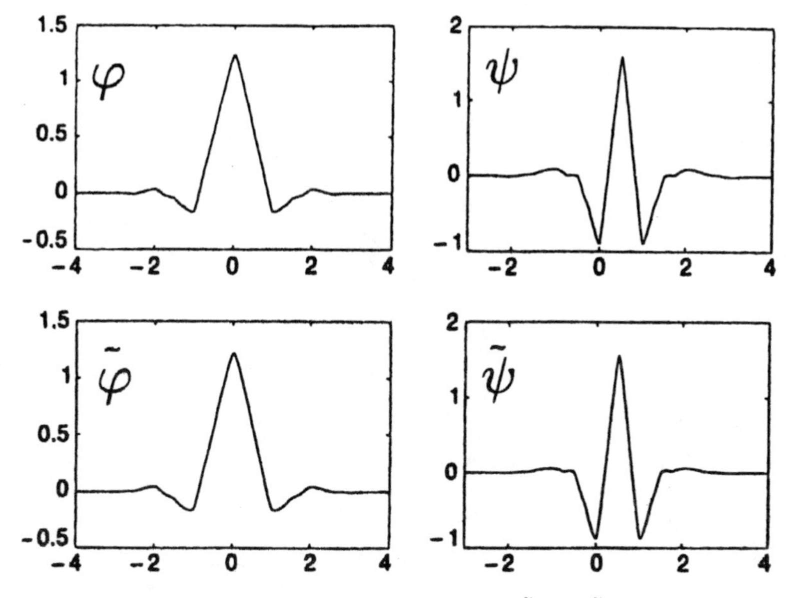

Figure 1.5. *Biorthogonal functions* ϕ, $\tilde{\phi}$, ψ, $\tilde{\psi}$ *for* $N = 2$.

Filter coefficients for Daubechies' symmetric, biorthogonal wavelets were taken from Table 8.5 in [7]. These wavelets were suggested by M. Barlaud and subsequently constructed by Daubechies. The filter coefficients for biorthogonal bases are very close to those of Coiflets; however, they are simpler to compute and yield perfectly symmetric wavelets. Graphs of the scaling functions φ, wavelets ψ, and their duals $\tilde{\varphi}$ and $\tilde{\psi}$, corresponding to $N = 2$, are shown in Figure 1.5.

1.3.4. Implementation of Mallat's Algorithm

We now describe how to implement Mallat's algorithm using filter coefficients for the scaling functions and wavelets described above. Data for a curve consist of a set of distinct points $\mathbf{P} = \{p_0, p_1, \ldots, p_n\}$. We follow the notation used in equation (1.5) and set $C_0^1 = x_0$, $C_1^1 = x_1$, $C_2^1 = x_2, \ldots$, $C_n^1 = x_n$, where $n = 2^k - 1$. Then we apply Mallat's algorithm to decompose the $\mathbf{C}^1$ into a set of coefficients $\mathbf{C}^0 = (C_0^0, C_1^0, \ldots, C_m^0)$ and $\mathbf{D}^0 = (D_0^0, D_1^0, \ldots, D_m^0)$ as in equations (1.5) and (1.6). Mallat's algorithm is, in turn, applied to the output reduced data set $\mathbf{C}^0 = (C_0^0, C_1^0, \ldots, C_m^0)$. We keep repeating the procedure so that, for the kth step, we apply the algorithm to the scaling function coefficients $\mathbf{C}^k$ to obtain $\mathbf{C}^{k-1}$ and $\mathbf{D}^{k-1}$. We repeat the process as many times as possible to reduce the data, while

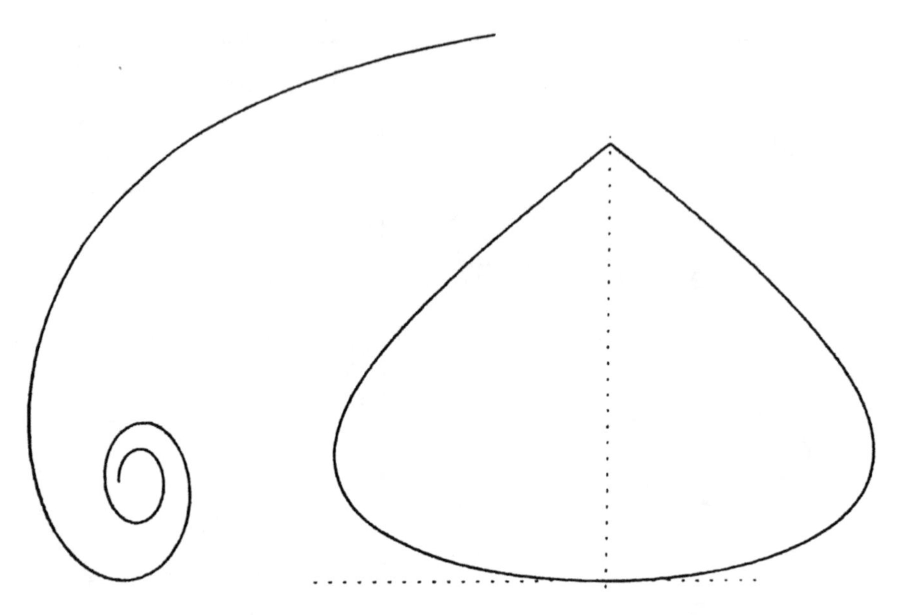

Figure 1.6. *Original spiral and Kuri curves*—1024 *data points.*

retaining a desired degree of accuracy of a curve reconstructed from the coefficients from a coarser level. Whenever a coefficient is very small, we replace it with zero to obtain further reduction of data.

In our discussions below, we use the following shorthand notation to represent the wavelets used in our experiments.

D_n	...	Daubechies' orthonormal wavelet;
C_n	...	Coiflets;
$S_{\tilde{n},n}$	...	Spline biorthogonal wavelet;
B_n	...	Daubechies' biorthogonal wavelet.

Experiment 1: Approximation of Spiral Curves

In our first set of experiments, we examined which wavelets would be best for approximating spiral curves. Spiral curves can be expressed parametrically as

$$x = (1/\theta)\cos\theta, \qquad y = (1/\theta)\sin\theta,$$

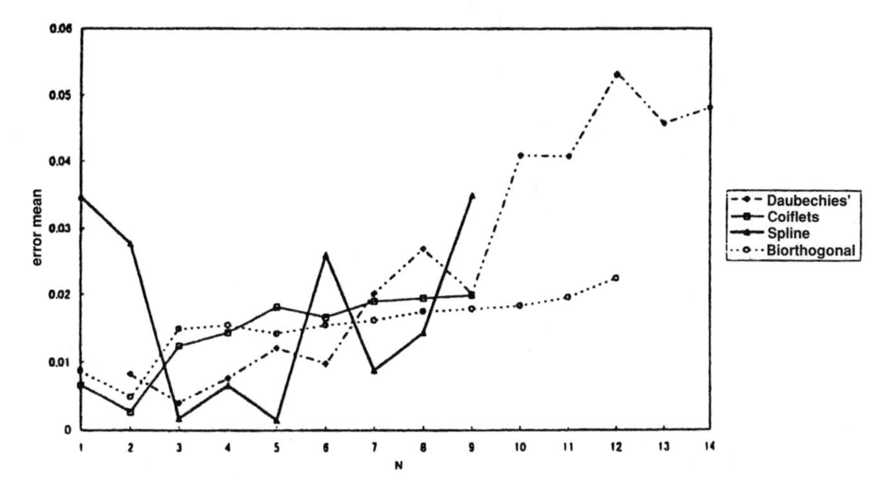

Figure 1.7. *Error in approximation of spiral using* 80 *points.*

where $0.2\pi \leq \theta < 5\pi$. We set the gridsize to $\Delta\theta = 4.8\pi/1024$ to obtain data consisting of 1024 points (see Figure 1.6) then apply Mallat's algorithm to reduce the data to 80 points, i.e., we constructed an approximation from 80 wavelet functions.

The error in the approximation is determined as follows. Let $P = \{p_0, p_1, \ldots, p_n\}$ denote the data points on the original curve and $P' = \{p'_1, \ldots, p'_n\}$ the data for the approximation to the curve. Then the errors are computed from

$$\text{error} = \sqrt{\sum_i d_i/n},$$

where d_i denotes the distance in the L^2-norm from the point p_i on the original curve to the corresponding point p' on the approximation. A graph illustrating the error, with respect to the type of wavelet and the degree of the filter coefficients N, is given in Figure 1.7. The wavelet bases used in our experiments are shown in Table 1.3. We concede that small values for the error in the L^2-norm may not always correspond to a good approximation; however, it is one quantitative measure of the quality of an approximation; as with most curve, graphics, and display problems, the best assessment is qualitative, so that drawings of the approximation should be displayed together with the original curve, as in Figure 1.8.

Table 1.3. *Wavelets and the degree of their filter coefficients.*

Wavelet	Degree
D_n	$2, 3, \ldots, 14$
C_n	$1, 2, \ldots, 9$
$S_{\tilde{n},n}$	$\tilde{N} = 1, \quad N = 1, 3, 5, 7, 9$
	$\tilde{N} = 2, \quad N = 2, 4, 6, 8, 10$
	$\tilde{N} = 9, \quad N = 9, 11, 13$
B_n	$1, 2, \ldots, 12$

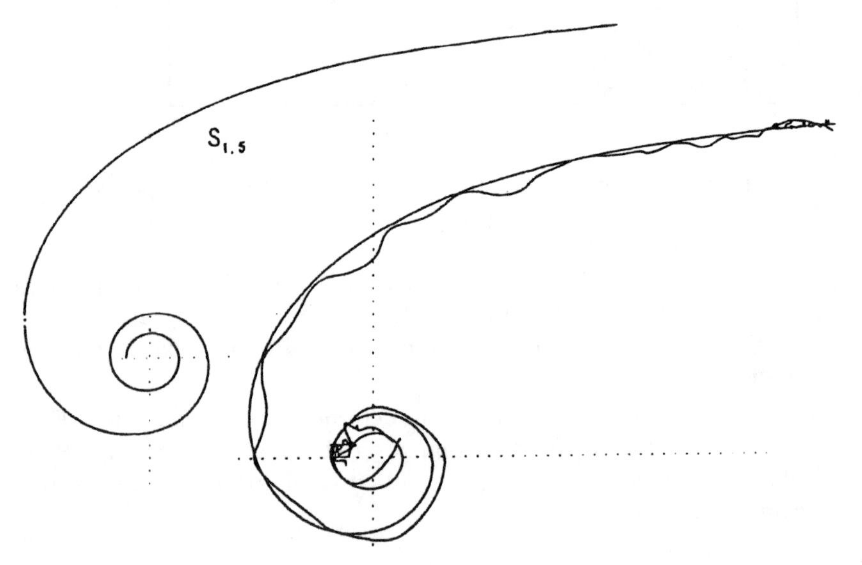

Figure 1.8. *The best and a poor approximation using* $S_{1,5}$ *and* $S_{3,9}$.

Based on our experimental results using the four types of wavelets described above, we conclude that the best results from each of the different types of wavelets were C_2, B_2, D_3, and $S_{1,3}$; that biorthogonal spline wavelets were the most sensitive to changes in the degree; that Daubechies' orthonormal wavelets yielded less accurate results as the degree of the wavelets used for reconstruction increased; and that results from using Coiflets and Daubechies' biorthogonal wavelets were stable with respect to changes in the degree of the filter coefficients. Approximations from the best cases for each type of wavelet coincide perfectly with the original curve for the resolution of our computer screen and printer, i.e., the graphs over-

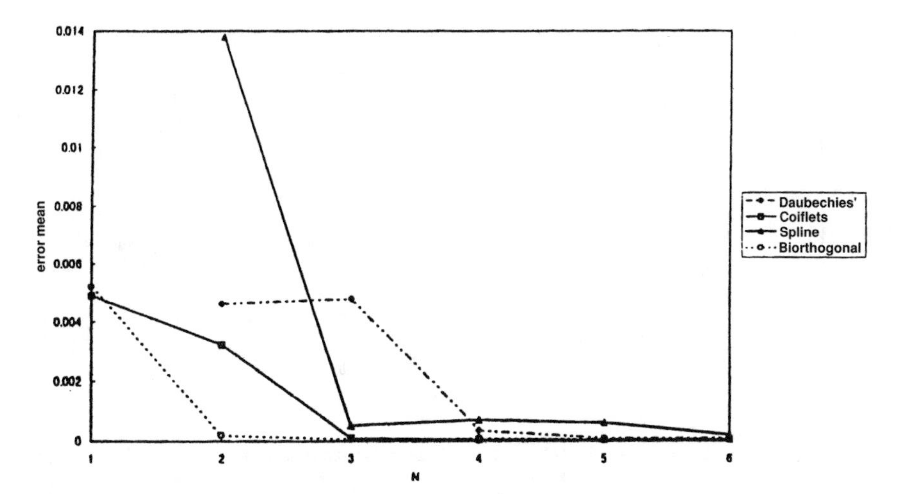

Figure 1.9. *Error in approximation of Kuri curve using* 60 *points.*

lap perfectly. One of the best examples, $S_{1,5}$, and an inaccurate example $S_{3,9}$, are shown in Figure 1.8.

Experiment 2: Approximation of Kuri Curves

In a second set of experiments, we examined which wavelets would be best for approximating Kuri curves. Kuri curves can be expressed parametrically as

$$x = \sin(2t), \qquad y = 1 - \cos(t),$$

where $-\pi/2 \leq t \leq \pi/2$. We set the gridsize to $\pi/1024$ and began with data consisting of 1024 points (see Figure 1.6).

We examined two cases—reduction of the data to 60 points and 40 points—and determined the errors following the same procedure as in the spiral curve experiments. Our results are plotted in Figures 1.9 and 1.10, respectively. We changed the compression ratio as well as the degree of the filter coefficients. Based on our results, we conclude that, in both sets of experiments, i.e., reduction to 60 and 40 points, the best results for each of the last three wavelet types were obtained using $S_{1,5}$, B_2, and C_6. And for all three wavelets, the errors did not change much as the degree increased, particularly for spline wavelets. For Daubechies' orthonormal wavelets, the

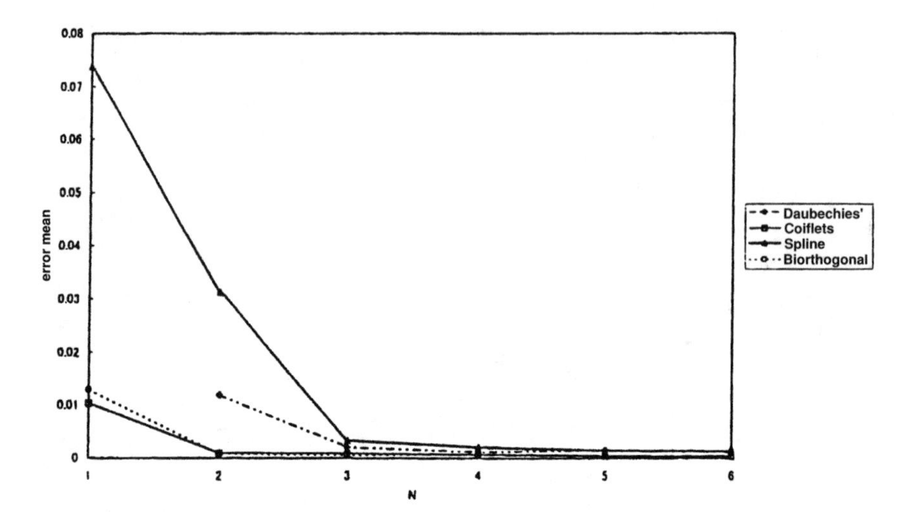

Figure 1.10. *Error in approximation of Kuri curve using* 40 *points.*

best results when reducing data to 60 points were for $N = 3$, and $N = 4$ when reducing data to 40 points. We also found that, in general, reducing data leads to greater loss of accuracy in the approximation, both quantitatively and qualitatively. However, there were some exceptions, e.g., reducing data to 40 points using D_3 wavelets yielded more accurate approximations than reducing data to 60 points using D_2 wavelets.

Experiment 3: Approximation of Coastlines

In a third set of experiments, we examined which wavelets would be best for approximating coastline curves. Coastlines have smooth and zig-zag features, both of which must be preserved (see Figure 1.11). We examined the effect of reducing data consisting of 256 points to 128 points and 64 points. The quantitative errors given in Figures 1.12 and 1.14 were computed in the same manner as in earlier experiments with spirals and kuri curves. A qualitative assessment can be made from Figures 1.13 and 1.15.

Based on our experimental results to reduce data to 128 points, we conclude that the best approximations for each type of wavelet were obtained using D_4, C_2, $S_{1,3}$, and B_2. Approximations from using Daubechies' orthonormal wavelets yield curves which are inaccurate, particularly in regions where the original is very smooth. Use of biorthogonal spline wavelets leads to approximations which are very sensitive to changes in the degree of

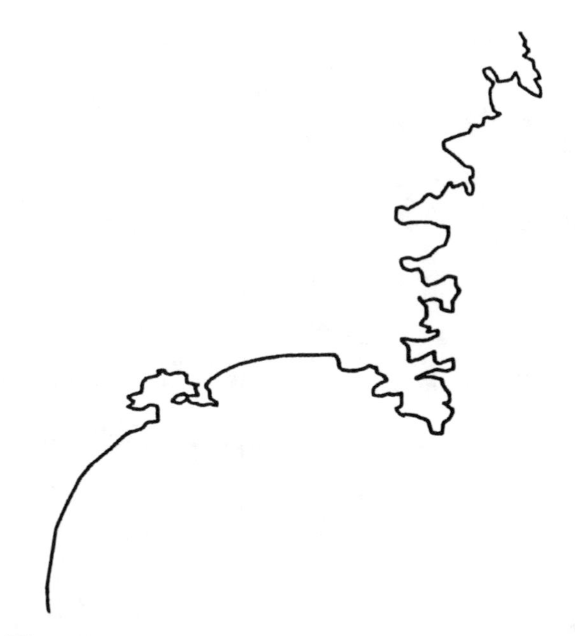

Figure 1.11. *Original coastline—256 data points.*

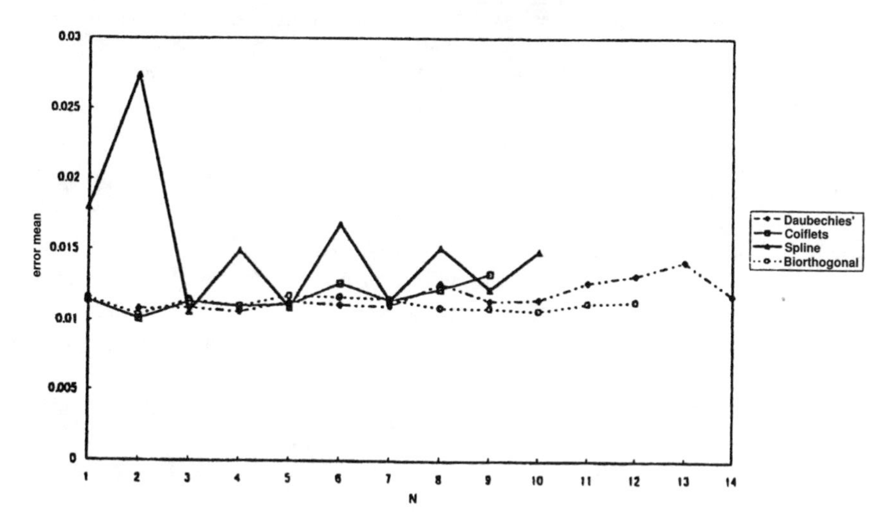

Figure 1.12. *Error in approximation of coastline using* 128 *points.*

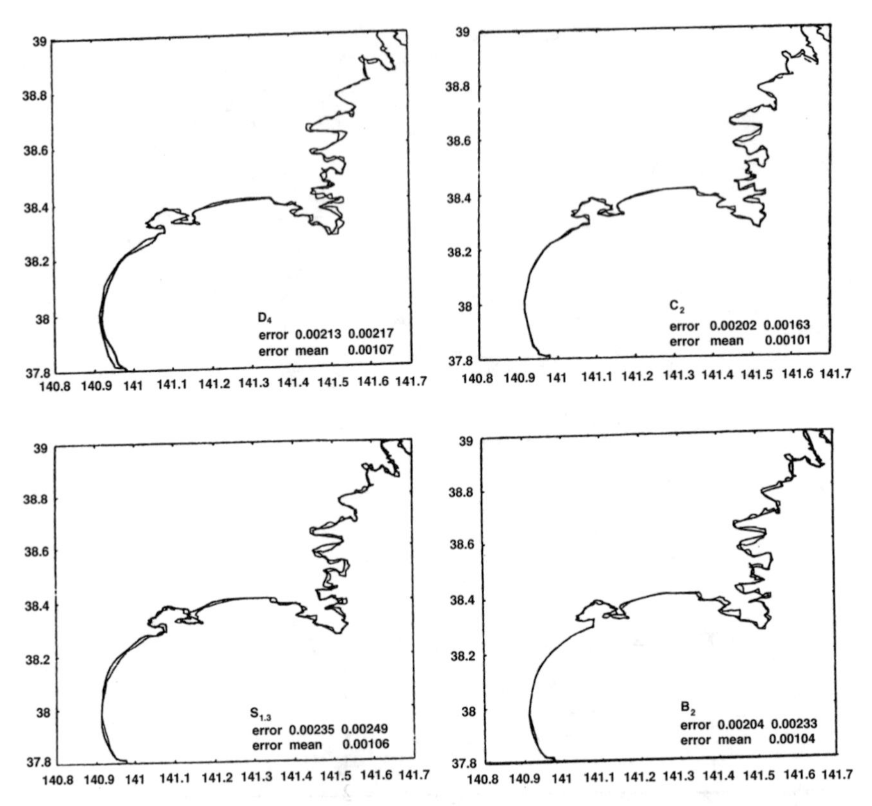

Figure 1.13. *Best approximations of coastline using* 128 *points.*

the wavelet and are too smooth in regions where the coastline has intricate features. Spline-based approximations are too inaccurate and unsuitable for coastline approximation because they sometimes yield curves which intersect themselves. Results obtained from using Coiflets and Daubechies' biorthogonal wavelets are stable, with respect to changes in the degree of filter coefficients, and use of Coiflets led to particularly well-fitting approximations to the original coastline curve.

In our experiments to reduce the data by a factor of 64, the best results for each particular type of wavelet were D_3, C_1, $S_{1,3}$, and D_1. The approximations from using Daubechies' and spline wavelets were sensitive to changes in degree; for higher degrees, Daubechies' wavelets lead to approximations which have too many zig-zags and have splines and also to approximations which are too smooth. Both Coiflets and biorthogonal wavelet-based approximations are stable with respect to changes in degree.

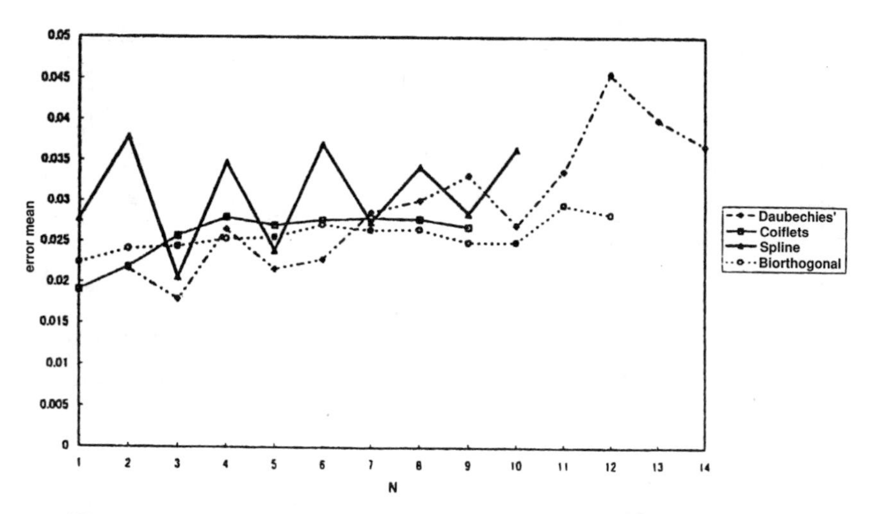

Figure 1.14. *Error in approximation of coastline using* 64 *points.*

The experiments described in this paper were conducted to assess the viability of wavelet-based multiresolution methods for displaying coastline data. Our results indicate that not all wavelet types can be used. In particular, we found that Coiflets and Daubechies' symmetric biorthogonal wavelets yield excellent approximations to coastline curves. The experiments described above use fewer data points than are available to ERI scientists. We are currently installing over 20,000 data points of the Izu coastline into the database of our system.

1.4. Conclusion

In this chapter we presented results from experiments to determine wavelets which are suitable for approximating two-dimensional coastline data at various resolution levels. In two preliminary studies to approximate spiral and kuri curves, we found that use of wavelets which lack symmetry and have fractal-like features yielded poor results.

In follow-up experiments using Japanese coastline data, we found that the MRA algorithm introduced by Mallat is useful for reducing coastline data and accurately approximating curves which have both zig-zag and smooth components. In particular, Coiflets and biorthogonal wavelets preserve the symmetries which are present in the original curve and yield consistently good results, with changes in the degree. Using biorthogonal wavelets with $N = 2$ yielded the best approximation. The MRA algorithm is very effective in reducing the amount of data needed to accurately display

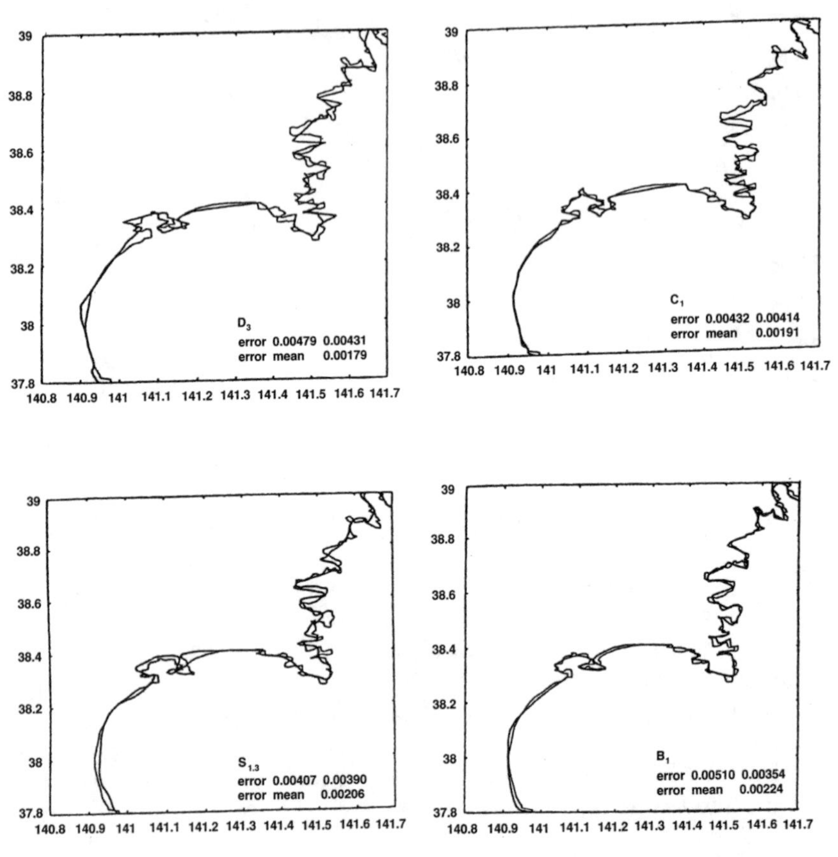

Figure 1.15. *Best approximations of coastline using* 64 *points.*

coastline data for a specified degree of resolution. However, one should be careful to select appropriate wavelets. When we choose inappropriate degrees, the accuracy falls even for small compression ratios, e.g., Daubechies' orthonormal wavelets and spline wavelets of high degree. Biorthogonal spline wavelets are only accurate for approximating very smooth curves.

The multiresolution techniques used in our experiments can be extended, more generally, for use in free-form curve design systems. As in the problem we considered, the wavelet type is likely to be an important factor in determining the accuracy of the results. Use of preliminary studies is recommended to identify important curve features for the problem at hand and to examine which wavelet types preserve these features.

More sophisticated methods for displaying three-dimensional views of geographic data are being developed by graphics experts; however, the methods require powerful computers with large memories and are not needed for the solution of our problem, i.e., to display coastlines with tens of thousands of points on a small workstation or PC in real time. As computers become faster and more powerful, even more sophisticated means of visualizing different types of geographic data on small workstations can and should be considered.

Acknowledgements. We gratefully acknowledge Kazuo Hatano's generosity in sharing his extensive tables of wavelet coefficients with us before publication of his manuscript, and we acknowledge Lena Marie Reissell for sending us preprints of her work which is closely related to ours.

References

[1] T. Agui, K. Miyata, M. Nakajima (1985), "Qosi-coding method for digital figures based on fractal dimension," *J. Inst. Television Engineers Japan*, vol. 39, pp. 979–987.

[2] M. Barnsley (1988), *Fractals Everywhere*, Academic Press, Tokyo.

[3] B. Barsky (1988), *Computer Graphics and Geometric Modeling Using Beta-splines*, Springer-Verlag, Tokyo.

[4] B. Bartles, J. Beatty, B. Barsky (1987), *An Introduction to Splines for Use in Computer Graphics and Geometric Modeling*, Morgan-Kaufmann, San Francisco, CA.

[5] A. Certain et al. (1996), "Interactive multiresolution surface viewing," *Proc. ACM Siggraph*, ACM, New York, pp. 91–98.

[6] I. Daubechies (1988), "Orthonormal bases of compactly supported wavelets," *Comm. Pure Appl. Math.*, vol. 41, pp. 909–996.

[7] I. Daubechies (1992), *Ten Lectures on Wavelets*, SIAM, Philadelphia, PA.

[8] I. Daubechies (1993), "Orthonormal bases of compactly supported wavelets II," *SIAM J. Math. Anal.*, vol. 24, pp. 499–519.

[9] D. Douglas, T. Peucker (1973), "Algorithms for the reduction of the number of points required to represent a digitized line or its caricature," *Canad. Cartographer*, vol. 10, pp. 112–122.

[10] M. Eck, H. Hoppe (1996), "Automatic reconstruction of B-spline surfaces of arbitrary topological type," *Proc. ACM Siggraph*, ACM, New York, pp. 325–334.

[11] G. Farin (1990), *Curves and Surfaces for Computer Aided Geometric Design*, Academic Press, Tokyo.

[12] A. Finkelstein, D. Salesin (1994), "Multiresolution curves," *Proc. ACM Siggraph*, ACM, New York, pp. 261–268.

[13] M. Gross, O. Staadt, R. Gatti (1996), "Efficient triangular surface approximation using wavelets and quadtree data structures," *IEEE Trans. on Visualization and Computer Graphics*, vol. 2, pp. 130–143.

[14] K. Hatano (1995), Private communication, Kyoto Univ. RIMS Workshop (http://phase.etl.go.jp/contrib/wavelet/).

[15] C. Herley (1993), *Wavelets and Filter Banks*, Ph.D. Thesis, Columbia Univ., New York, NY.

[16] S. Hiyama (1994), "The data reduction of plane curves," *Proc. 3rd Intl. Colloquium on Numerical Analysis*, Faculty of Mathematics, Ploveliv Univ., Bulgaria, pp. 77–86.

[17] S. Hiyama, T. Hanada, H. Imai (1990), "An optimum data reduction algorithm for general plane curves," *Technical Report* ISE-TR-90-87, Inst. of Information Science and Electronics, Univ. of Tsukuba, Japan.

[18] S. Hiyama, T. Hanada, H. Imai (1996), "The data reduction of a shoreline by wavelet transforms," *Trans. Japan Soc. Indust. Appl. Math.*, vol. 6, pp. 83–99 (in Japanese).

[19] H. Hoppe (1996), "Progressive meshes," *Proc. ACM Siggraph*, ACM, New York, pp. 99–108.

[20] H. Hoppe (1997), "View-dependent refinement of progressive meshes," *Proc. ACM Siggraph*, ACM, New York, pp. 189–198.

[21] H. Imai, M. Iri (1986), "Computational-geometric methods for polygonal approximation of a curve," *Computer, Graphics and Image Processing*, vol. 36, pp. 31–41.

[22] P. Lindstrom et al. (1996), "Real-time, continuous level of detail rendering of height fields," *Proc. ACM Siggraph*, ACM, New York, pp. 109–118.

[23] D. Luebke, C. Erikson (1997), "View-dependent simplification of arbitrary polygonal environments," *Proc. ACM Siggraph*, ACM, New York, pp. 199–208.

[24] S. Mallat (1989), "Multiresolution approximation and wavelets," *Trans. Amer. Math. Soc.*, vol. 315, pp. 69–88.

[25] B. Mandelbrot (1982), *The Fractal Geometry of Nature*, W. H. Freeman, San Francisco, CA.

[26] R. McMaster (1989), "The integration of simplification and smoothing algorithms in line generation," *Cartographica*, vol. 26, pp. 101–121.

[27] L.-M. Reissell (1993), "Multiresolution geometric algorithms using wavelets I: Representation for parametric curves and surface," *Technical Report* TR-17, Dept. Comput. Sci., Univ. of British Columbia, Vancouver, Canada.

[28] O. Rioul, M. Vetterli (1991), "Wavelets and signal processing," *IEEE Trans. Signal Process.*, Oct., pp. 14–38.

[29] E. Shikin, A. Plis (1995), *Handbook on Splines for the User*, CRC Press, Tokyo.

[30] E. Stollnitz, T. DeRose, D. Salesin (1996), *Wavelets for Computer Graphics*, Morgan-Kaufmann, San Francisco, CA.

[31] G. Strang, T. Nguyen (1996), *Wavelets and Filter Banks*, Wellesley–Cambridge Press, Wellesley, MA.

[32] Ph. Tchamitchian (1987), "Biorthogonalité et théorie des opérateurs," *Rev. Mat. Iberoamericana*, vol. 3, pp. 163–189.

[33] P. Vaidyanathan (1992), *Multirate Systems and Filter Banks*, Prentice–Hall, Englewood Cliffs, NJ.

[34] M. Vetterli, C. Herley (1992), "Wavelets and filter banks: Theory and design," *IEEE Trans. Signal Process.*, vol. 40, pp. 2207–2232.

[35] M. Vetterli, J. Kovačević (1995), *Wavelets and Subband Coding*, Prentice–Hall, Englewood Cliffs, NJ.

[36] H. William, A. Saul, T. William, P. Brian (1992), *Numerical Recipes in C*, 2nd ed., Cambridge Univ. Press, Cambridge, pp. 591–608.

[37] D. Zorin, P. Schröder, W. Sweldens (1996), "Interpolating subdivision for meshes with arbitrary topology," *Proc. ACM Siggraph*, ACM, New York, pp. 189–192.

[38] D. Zorin, P. Schröder, W. Sweldens (1997), "Interactive multiresolution mesh editing," *Proc. ACM Siggraph*, ACM, New York, pp. 259–268.

2. A Wavelet-Based Technique for Reducing Noise in Laboratory Data

Susumu Sakakibara*

Abstract. This chapter presents results from a case study on a very simple and inexpensive wavelet-based technique to reduce noise in experimental data, rendering it clean enough to be of practical value. We reduce noise by using a decomposition algorithm of Mallat's and subtracting higher level components of the wavelet expansion of the data. The cleaner data can be used to compute the velocity accurately using finite differences of the displacement; finite difference methods augment noise so that they cannot be used on noisy displacement data. We present two examples, dry friction analysis and a drop mass test, to demonstrate how our method yields cleaner data which facilitate the study of mechanical systems.

Key words. wavelet, noise reduction, dry friction analysis, drop mass test

2.1. Introduction

In the experimental study of dynamics of mechanical systems, scientists are interested in estimating the displacement and velocity of components of a system. The displacement data are a set of discretely sampled points which are processed and recorded by a computer connected to measurement devices. The velocity, which is defined as the temporal derivative of the displacement, can be approximated using finite differences of the discretely sampled displacement data. However, finite differences are very sensitive to noise so that velocity estimates are often too inaccurate to be of practical value. In this chapter, we examine how wavelet analysis can be used to eliminate noise and find smooth functions which accurately approximate the displacement and velocity. Our method has several advantages compared to conventional Fourier or polynomial fitting methods: the associated computations are simple and inexpensive, and since the wavelets we

*College of Science and Engineering, Iwaki Meisei University, Iwaki-shi, Fukushima-ken 970-8551 Japan (susumu@iwakimu.ac.jp).

consider have compact support, they are well suited for processing signals of finite duration.

Our noise reduction method is based on the assumption that the displacement and velocity functions are reasonably smooth. Wavelet components of higher levels are discarded to reduce noise. Through two experimental examples, we demonstrate how a very simple and inexpensive wavelet-based noise reduction method is useful in engineering applications and discuss how to determine the level above which the components should be eliminated. In the examples, smoothed displacement and velocity signals are used to determine properties of a mechanical system. The primary merit of our method is simplicity in implementation. The software we use is simple to implement, uses little memory, and can run on a simple notebook PC. More sophisticated and expensive approaches to wavelet-based noise analysis discussed in [6], [9], [11], [12], [14] and Chapter 5 of this book are not required to solve our problem.

This chapter is organized as follows. In the next section we establish our notation and describe our noise reduction method. The method is based on the decomposition algorithm of Mallat [13]. Properties of two wavelets (orthonormal wavelets with compact support [7] and cardinal B-splines [5]) used in our experiments are described. In the third section, we present two applications (dry friction analysis and a drop mass test) and demonstrate how meaningful, cleaner data can be extracted. Our findings are summarized in the fourth and final section.

2.2. Wavelets and a Data Smoothing Method

In this section we present a simple and effective wavelet-based noise reduction method for processing data from mechanical systems and describe properties of two wavelets used in our experiments (orthonormal wavelets with compact support and B-spline wavelets).

2.2.1. A Wavelet-Based Data Smoothing Method

Our wavelet-based data smoothing procedure is as follows. Suppose we are given data consisting of a finite set of values of the displacement $\{x_n\}$; $n \in \mathbf{Z}$ sampled at evenly spaced time intervals of length Δt. Our goal is to determine a smooth function

$$f_j(t) = \sum_{k \in \mathbf{Z}} c_k^{(j)} \, \phi(2^j t - k), \qquad j \in \mathbf{Z}, \tag{2.1}$$

which approximates the data, i.e., $f_j(n\Delta t) \sim x_n$. To simplify notation, we set $\Delta t = 1$ so that the approximation satisfies $f_0(n) \sim x_n$. Mallat set the initial coefficients to be $c_k^{(0)} \approx x_k$ in an image processing application [13]; however, this choice is not suitable for our application. We discuss a better initialization technique based on B-splines later in this section.

We use an algorithm of Mallat [13] and decompose the approximation (2.1) into coarse and detailed components $\{c_k^{(j)}\}$ and $\{d_k^{(j)}\}$; $j = -1, -2, \ldots,$ $k \in \mathbf{Z}$, where

$$c_k^{(j-1)} = \frac{1}{2} \sum_{\ell} \overline{p}_{\ell-2k}\, c_\ell^{(j)} ,$$

$$d_k^{(j-1)} = \frac{1}{2} \sum_{\ell} \overline{q}_{\ell-2k}\, c_\ell^{(j)} ,$$

and $\{p_k\}$, $\{q_k\}$; $k \in \mathbf{Z}$ are filter coefficients of the scaling function $\phi(t)$ and wavelets $\psi(t)$ in the dilation equations

$$\phi(t) = \sum_k p_k\, \phi(2t-k) , \tag{2.2}$$

$$\psi(t) = \sum_k q_k\, \phi(2t-k) .$$

The p_k are normalized to satisfy $\sum_k p_k = 2$. Orthogonal wavelets satisfy the additional property $q_k = (-1)^k p_{1-k}$. The general form for the scaling function and wavelet expansion of $f_0(t)$ is

$$f_0(t) = f_J(t) + \sum_{j=-1}^{J} g_j(t) , \tag{2.3}$$

where

$$g_j(t) = \sum_k d_k^{(j)}\, \psi(2^j t - k) .$$

To reduce noise, we discard the detailed components $g_j(t)$ for $j \geq J$; $j \in \mathbf{Z}$. The smooth function

$$f_J(t) = f_0(t) - \sum_{j>J} g_j(t) \tag{2.4}$$

estimates the displacement $x(t)$ with reduced noise, and it can be used to explore properties of a system and to determine the velocity. The cut-off level J is determined from scientific experience and known properties of the resulting smooth function. Our choice of J is very much application dependent. The velocity is computed straightforwardly as the temporal derivative of the displacement $\dot{x}(t) \equiv dx/dt$, i.e.,

$$\dot{f}_j(t) = \sum_k c_k^{(j)} \, 2^j \, \dot{\phi}(2^j t - k) \; .$$

In the next subsections, we examine two wavelets which are suitable for use in our noise reduction method.

2.2.2. Daubechies' Wavelets

The scaling functions ϕ for orthonormal wavelets with compact support, or Daubechies' wavelets, are computed by iteration from their filter coefficients p_k and the equation

$$\phi(n) = \lim_{j \to \infty} \phi_j(n) \; ,$$

where

$$\phi_j(n) = \sum_k p_k \, \phi_{j-1}(2n - k)$$

[7]. The characteristic function is normally used as the initial function $\phi_0(n)$. As a consequence of the orthogonality property, the scaling function and wavelet filter coefficients p_k and q_k satisfy

$$q_k = (-1)^k \, p_{1-k}$$

and

$$\sum_n \phi(n) = 1 \; .$$

The derivative of the scaling function $\dot{\phi}(t)$ is computed by differentiating the dilation equation (2.2), i.e.,

$$\dot{\phi}(t) = \sum_k 2 \, p_k \, \dot{\phi}(2t - k) \; .$$

We set $t = n$; $n \in \mathbf{Z}$ to obtain

$$\dot{\phi}(n) = \sum_k 2 \, p_k \, \dot{\phi}(2n - k)$$

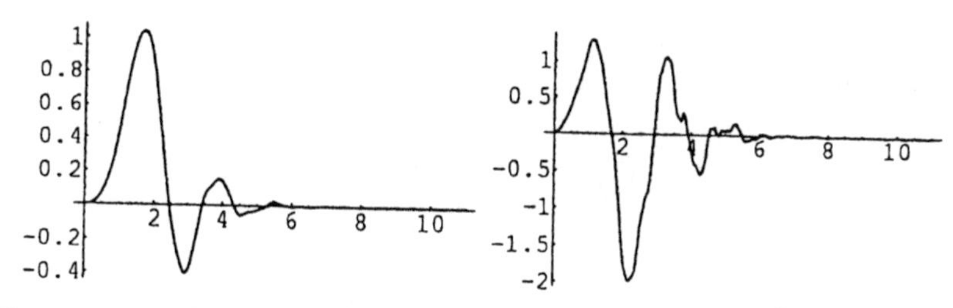

Figure 2.1. *The scaling function $\phi(t)$ and its derivative $\dot{\phi}(t)$ for $N = 6$.*

and the sum rule

$$\sum_n n \, \dot{\phi}(n) \;=\; -1 \, .$$

The derivative of the scaling function $\phi(t)$ is well behaved only when the number of vanishing moments N for the corresponding wavelet, defined as

$$\int_{-\infty}^{\infty} t^k \, \psi(t) \, dt \;=\; 0, \qquad k = 0, 1, \ldots, N-1,$$

is large. For example, $\phi(t)$ for $N = 2$ does not have a continuous derivative. Figure 2.1 shows $\phi(t)$ and $\dot{\phi}(t)$ for $N = 6$.

Although Daubechies' wavelets are orthonormal and convenient for computation, other properties offset their attractiveness for our application. For example, the values of $\phi(t)$ are very close to 0 in the right half of the interval of its support so that half of the function is not effectively used in representing $f_j(t)$ in (2.1). Furthermore, $\phi(t)$ and $\dot{\phi}(t)$ oscillate with relatively large amplitudes, with $\max|\dot{\phi}(t)|$ being twice as large as $\max|\phi(t)|$. These oscillations will appear in the approximations for $x(t)$ and $\dot{x}(t)$.

2.2.3. B-Spline Wavelets

Cardinal B-splines serve as the scaling functions for cardinal B-spline wavelets. These splines have been used extensively in computer graphics [3], [4], [10], [18] and are defined as

$$N_m(t) = \frac{1}{(m-1)!} \sum_{k=0}^{m} (-1)^k \binom{m}{k} (t-k)_+^{m-1} \, ,$$

where m is the order, $x_+ = \max\{0, t\}$, and $N_1(t) = \chi_{[0,1)}(t)$ is the characteristic function on the semi-open unit interval. To determine filter coefficients for B-splines, we compute the Fourier transform of the scaling function

$$\hat{N}_m(\omega) = \left(\frac{1-e^{-i\omega}}{i\omega}\right)^m ,$$

and find its dilation equation in Fourier form, i.e.,

$$\hat{N}_m(\omega) = P_m(e^{-i\omega/2})\, \hat{N}_m\left(\frac{\omega}{2}\right) .$$

The polynomial $P_m(z)$ used to determine the filter coefficients p_k is

$$P_m(z) = \frac{1}{2} \sum_k p_k\, z^k = \left(\frac{1+z}{2}\right)^m .$$

To derive B-spline wavelets, let

$$E_{N_m}(e^{-i\omega}) = \sum_k |\hat{N}_m(\omega + 2\pi k)|^2 .$$

Then

$$E_{N_m}(z) = \sum_{k=-m+1}^{m-1} N_{2m}(m+k)\, z^k$$

is a trigonometric polynomial for $z = e^{-i\omega}$, with rational coefficients. Let

$$\begin{aligned} Q_m(z) &= \frac{1}{2} \sum_k q_k z^k \\ &= -z\, E_{N_m}(-z)\, \overline{P_m(-z)} , \end{aligned}$$

where $z = e^{-i\omega}$. The B-spline wavelet ψ is defined as the function with a Fourier transform satisfying

$$\hat{\psi}_m(\omega) = Q_m(e^{-i\omega/2})\, \hat{N}_m\left(\frac{\omega}{2}\right)$$

[5]. Note that the scaling function $N_m(t)$ is nonnegative, symmetric, and has derivative

$$\dot{N}_m(t) = N_{m-1}(t) - N_{m-1}(t-1) , \tag{2.5}$$

so the function

$$f_j(t) = \sum_k c_k^{(j)}\, N_m(2^j t - k)$$

has the derivative

$$\dot{f}_j(t) = 2^j \sum_k \left(c_k^{(j)} - c_{k-1}^{(j)}\right) N_{m-1}(2^j t - k) \ .$$

These properties are desirable for interpolation of discrete data and implementation of our noise reduction algorithm.

2.2.4. Initialization of the Decomposition Algorithm

Earlier in this section, we mentioned that Mallat initialized his decomposition algorithm by setting the coefficients $c_k^{(0)}$ in (2.1) to be $c_k(0) = x_k$. We use an initialization algorithm better suited for our application which uses B-spline interpolation. Let $\{\beta_k^{(m)}\}$; $k \in \mathbf{Z}$ be the sequence defined by

$$\sum_k \beta_k^{(m)} \ z^k = \left(\sum_k N_m \left[\frac{m}{2} + k\right] \ z^k \right)^{-1} .$$

Set

$$c_k^{(0)} = \sum_\ell \beta_{k-\ell+m/2} \ x_\ell \ .$$

Then

$$f_0(t) = \sum_k \ c_k^{(0)} \ N_m(t-k)$$

interpolates the input data $\{x_n\}$ so that $f_0(n) = x_n$ when n is even. In implementations, we approximate the infinite sum (2.3) by the finite sum (2.4) by setting $\beta_k^{(m)} = 0$ for all $|k| \geq k_{\max}$ [8], [15], [16], [17].

Although B-spline scaling functions and wavelets are not orthogonal, corresponding dual scaling functions and wavelets exist. Since we discard the higher level wavelet components in our method, we do not need to know the dual functions per se for implementations. Finally, we mention that Chui used an alternate interpolation method which uses piecewise polynomials of fixed degree [5].

2.3. Two Examples

In this section we show how our noise reduction technique can be successfully applied to process data from two laboratory experiments: dry friction analysis and a drop mass test.

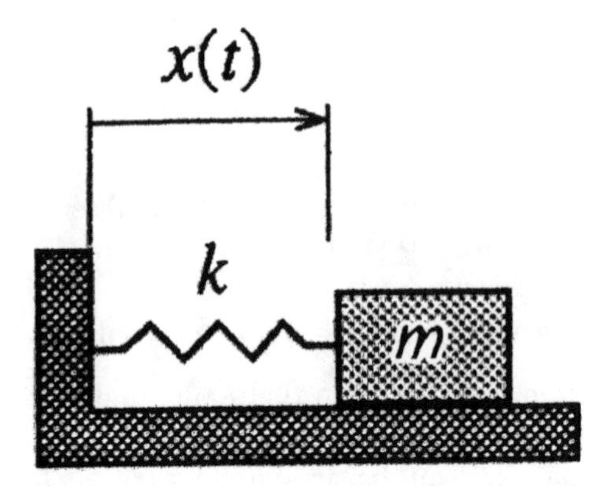

Figure 2.2. *Schematic model of vibration.*

2.3.1. Dry Friction Analysis

In our first experiment, to study properties of dry friction, a piping system of length 10 meters is subjected to an applied vibrational force to simulate an earthquake. The support of the system rests on a flat plate. When the support slides back and forth on the surface, the frictional force between the support and the plate dampens the vibrations of the system. The system (shown in Figure 2.2) is mathematically modeled by a one-dimensional equation of motion, where m is the mass of the piping system at the support and dry friction acts on the contact surface. The simplest model of friction is the Coulombic force model

$$f = -F_0 \operatorname{sign} v,$$

where v is the velocity of the support, and F_0 is a constant. We consider the modified model, shown on the left-hand side of Figure 2.3. Our goal is to study whether laboratory experimental data exhibit behavior predicted by this modified model.

Our experimental data consist of samples of the displacement $x(t)$ and the force $f(t)$ measured at evenly spaced time intervals $\Delta t = 0.005$ sec., i.e., $\{x_n\}$ and $\{f_n\}$; $n \in \mathbf{Z}$. We approximate the velocity by the central difference

$$v_n = \frac{x_{n+1} - x_{n-1}}{2\Delta t}$$

and produce a *vf*-plot with a very wide error band due to noise, shown on the right-hand side in Figure 2.3. The plots of x_n, v_n, and f_n in Figure 2.4 show that central differences augment the noise in the data.

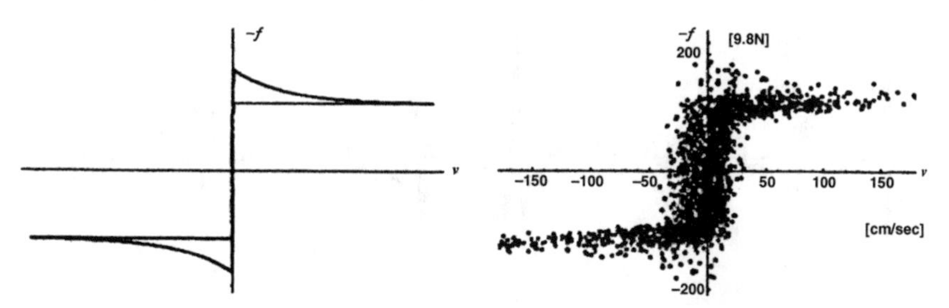

Figure 2.3. *Modified Coulombic force data (left) and corresponding laboratory data (right).*

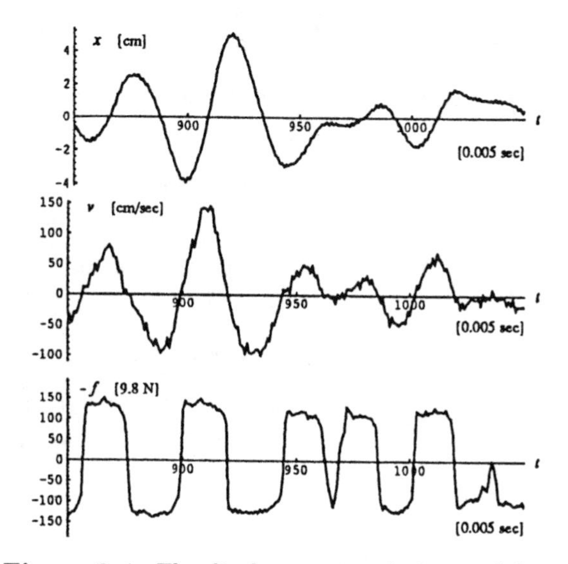

Figure 2.4. *The displacement, velocity, and force.*

Next, we consider a conventional low-pass digital filtering method to reduce noise using a Butterworth filter with transfer function

$$H(z) = \frac{b_1 z^{-1}}{1 + a_1 z^{-1} + a_2 z^{-2}} \ .$$

For a given input v_n, the output w_n satisfies

$$w_n = -a_1 w_{n-1} - a_2 w_{n-2} + b_1 v_{n-1} \ .$$

From analysis of laboratory data, we note that the fundamental mode is around 4–5 Hz, and noise is around 90 Hz and above. We set the cutoff

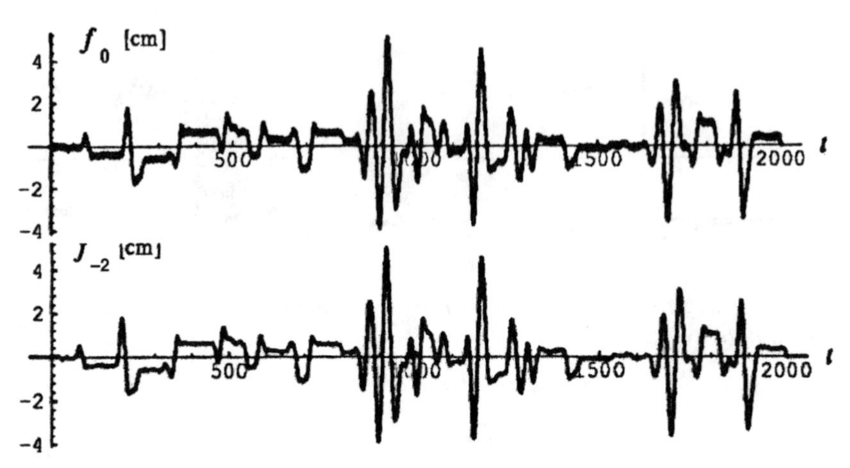

Figure 2.5. *Displacement data (top) and an estimate (bottom).*

frequency to be $f_c = 50$ Hz. The sampling frequency is $1/\Delta t = 200$ Hz so that

$$\begin{aligned} a_1 &= -0.292448, \\ a_2 &= +0.108453, \\ b_1 &= +0.816005. \end{aligned}$$

In Figure 2.8, we compare the graph of the velocity w_n (shown in the middle) with that obtained using finite differences (shown on the top).

In a third approach, we estimate the displacement $x(t)$ and the force $f(t)$ using $N_4(t)$ cubic spline interpolation on laboratory data. The initial coefficients $\{c_k^{(0)}\}$ for implementation of Mallat's decomposition algorithm are computed from the laboratory data as described in section 2.4. Results from the approximation of $f(t)$ are shown in Figure 2.5. The wavelet components $g_j(x)$ corresponding to the wavelet coefficients $\{d_k^{(j)}\}$ show that for $j = -1$ and -2, noise predominates, while for $j = -3$, important structural information is present (Figure 2.6). Based on this observation, we select the cutoff level to be $J = -2$ and use $f_{-2}(t)$ as our estimate of $x(t)$ (see bottom of Figure 2.5). Figure 2.7 illustrates the effectiveness of our wavelet-based method in smoothing the displacement data (compare with Figure 2.4).

When the velocity $\dot{x}(t)$ is computed using equation (2.5), and data is processed by our wavelet-based technique, our results are much better than those computed using finite difference and Butterworth filtering methods (Figure 2.8). Our wavelet method reduces the noise while preserving details of local fluctuations.

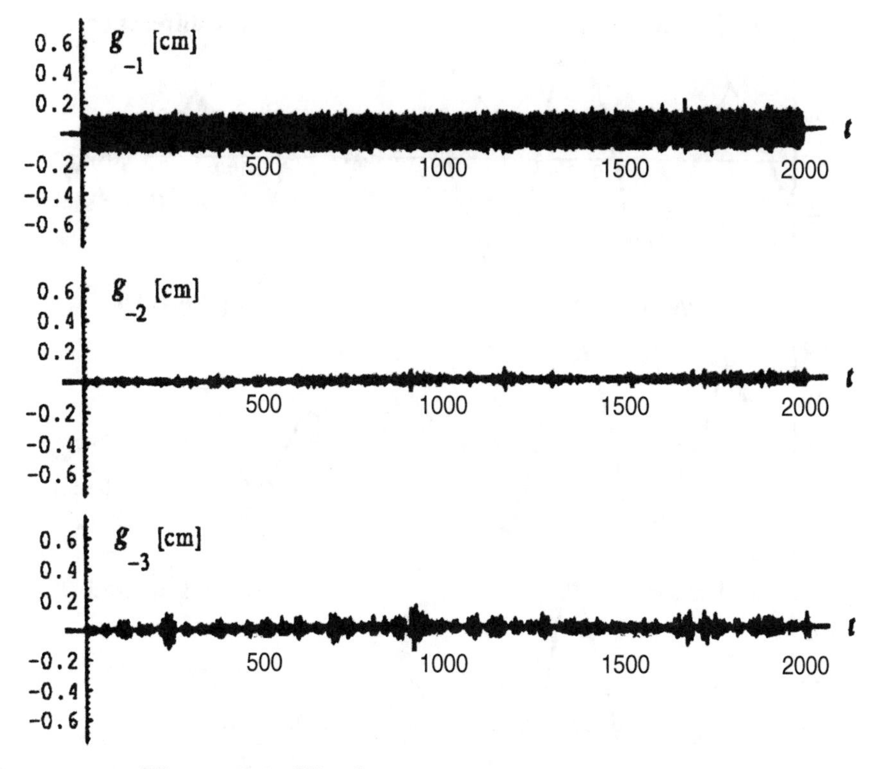

Figure 2.6. *Wavelet components* g_j; $j = -1, -2, -3$.

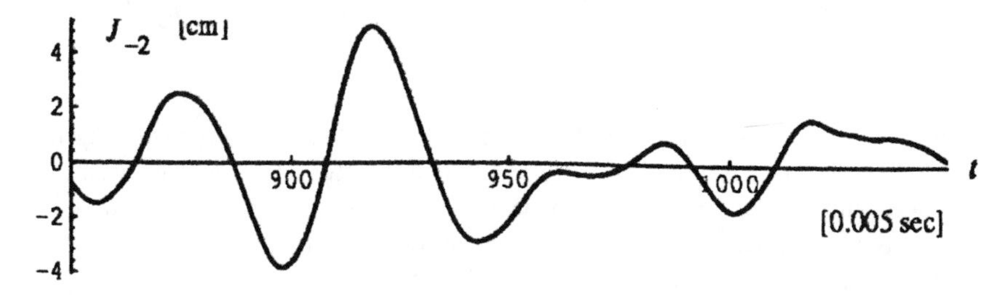

Figure 2.7. *A smoothed estimate for the displacement.*

Unfortunately, we were unsuccessful in reducing noise in the force data $\{f_n\}$ using wavelet methods. The frictional force $f(t)$ resembles a piecewise constant function, as can be seen in the bottom of Figure 2.4. Our method smooths the curve so that it does not retain its piecewise constant property.

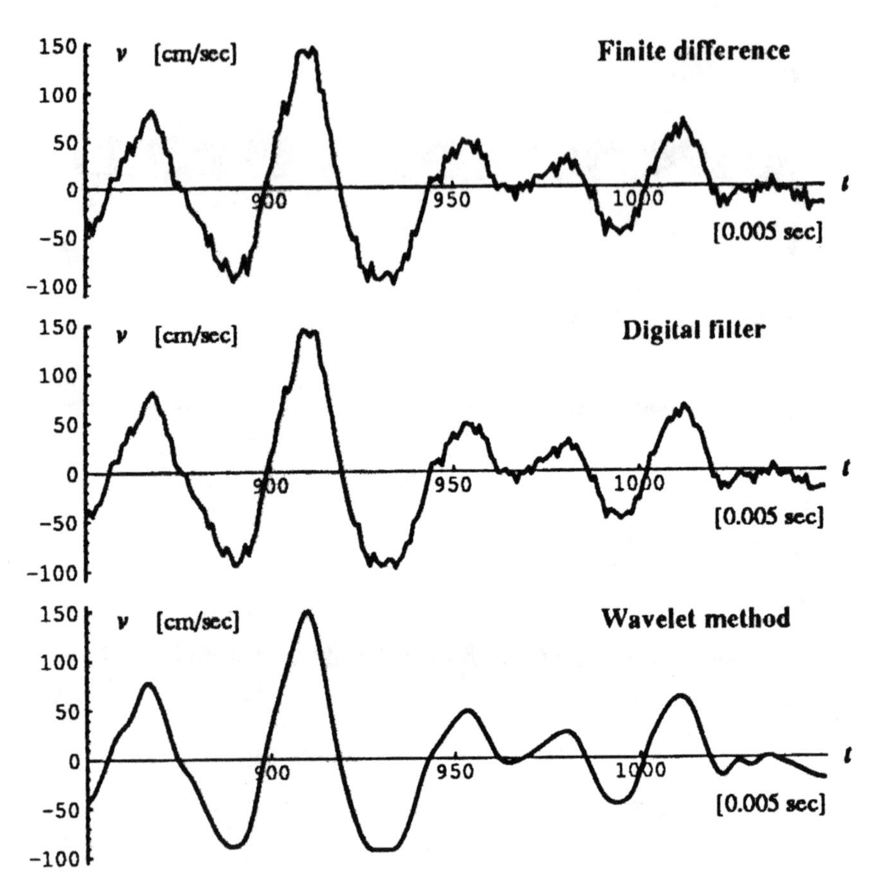

Figure 2.8. *Estimates for the velocity obtained from three methods.*

vf-plots in which only the displacement data is clean are given in Figure 2.9 for the conventional digital filter (shown on the left) and the wavelet method (shown on the right). Comparison with Figure 2.3 shows that both methods reduce the noise; however, the *vf*-plot on the right in Figure 2.9 is sharper and has a slimmer error band.

Although it is not possible to show the precise behavior of dry friction from our *vf*-plot, the increased sharpness of the graphs still provides valuable information. For example, we found several cycles that are close to predictions (see Figure 2.10). These cases show that the frictional force is small when the magnitude of velocity is small. In future experiments, further refinements of our setup will be made, which are expected to lead to even better results. For example, we plan to develop a more

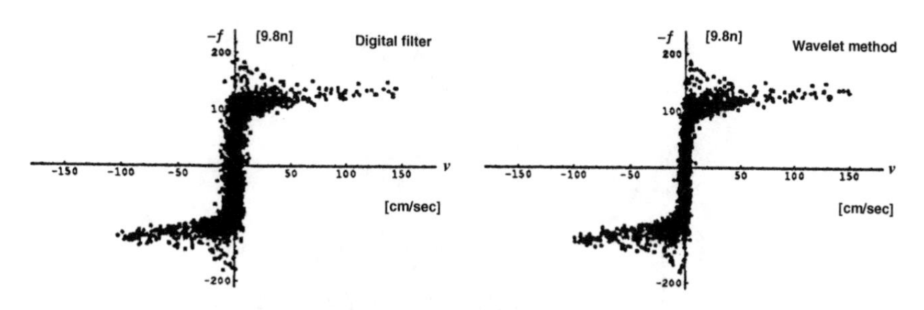

Figure 2.9. *vf-plots from digital filter (left) and wavelet methods (right).*

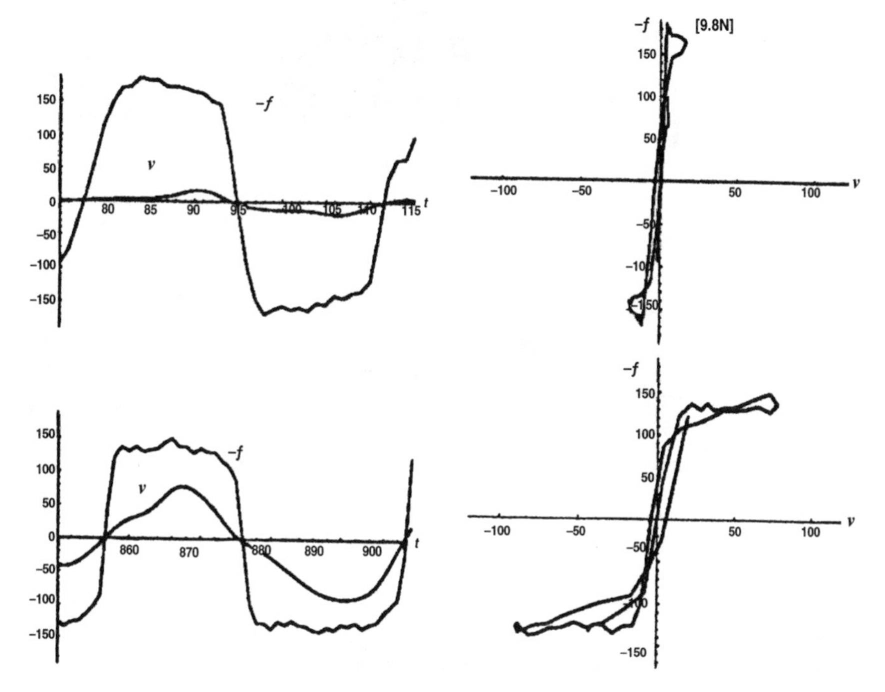

Figure 2.10. *Selected one cycles.*

sophisticated model to take into account that displacements are not perfectly one-dimensional movements, as assumed in our simple model above.

2.3.2. Drop Mass Test

In our second experiment, we performed a drop mass test to study the energy absorption properties of silicone. A weight of mass $m = 3$ kg was

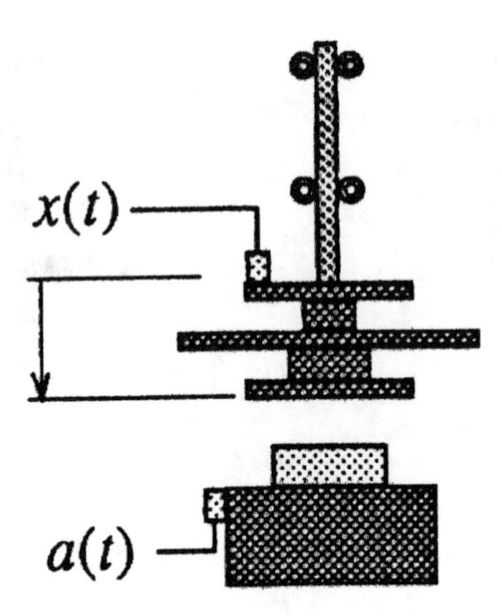

Figure 2.11. *Experimental setup of the drop mass test.*

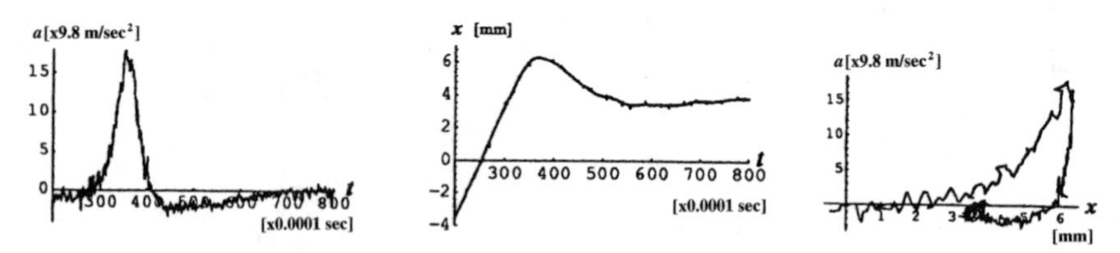

Figure 2.12. *Acceleration and displacement data.*

dropped on a small sheet of silicone gel. The vertical displacement $\{x_n\}$ and acceleration $\{a_n\}$; $k \in \mathbf{Z}$ of the weight were measured at equally spaced time intervals of $\Delta t = 0.0001$ sec. Initial heights of the weight were set at $h = 3$, 6, and 9 cm. We plot the points (x_n, a_n) on the x-a plane. Since ma_n represents the force acting on the mass, the area of the region enclosed by the curve times m is equal to the energy absorbed by the gel, i.e.,

$$\text{energy} = \int f\,dx = \int m\,a\,dx.$$

This formula represents only a part of the potential energy of the weight at the initial height, mgh; the rest is dissipated into the floor on which the apparatus is set.

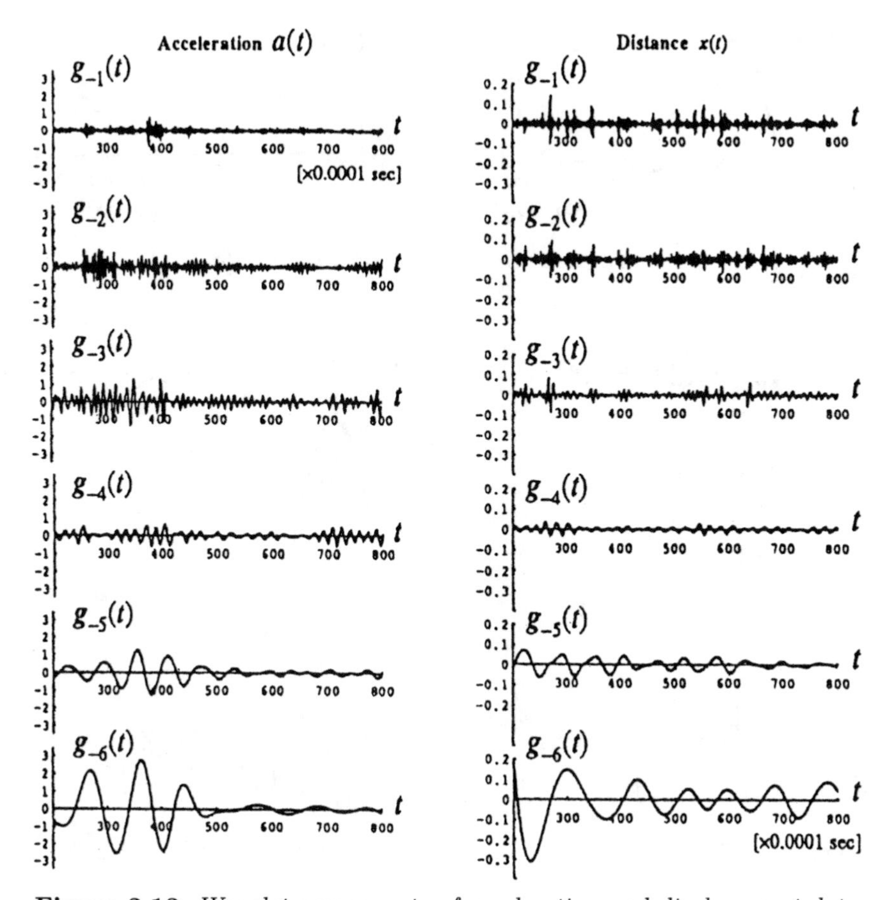

Figure 2.13. *Wavelet components of acceleration and displacement data.*

Our experimental scheme is illustrated in Figure 2.11. The weight is guided by roller bearings which introduce noise into the data. In theory, the acceleration is just the second derivative of the displacement, so we should be able to compute it from our displacement data; however, the noise is sufficiently large that separate measurements of the acceleration are needed. Our aim is to eliminate noise in our data and to determine the energy absorbed by the silicone gel.

We describe our method using data corresponding to $h = 3$ cm. The original data are used to generate plots of acceleration $\{a_n\}$, displacement $\{x_n\}$, and x-a as shown in Figure 2.12. To reduce the noise in the data, we compute the scaling function and wavelet coefficients $c_k^{(0)}$, $d_k^{(j)}$; $j = -1$, $-2, \ldots$ separately for the acceleration and displacement data. The wavelet

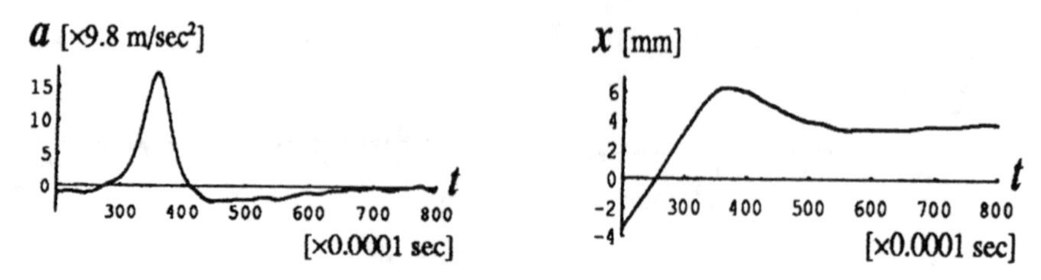

Figure 2.14. *Smoothed acceleration data.*

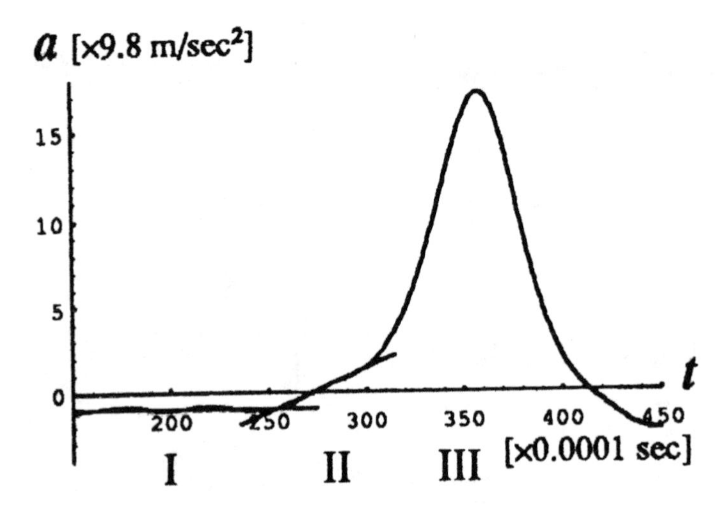

Figure 2.15. *Regions in which acceleration changes linearly* (*in* I *and* II) *and continuously* (*in* III).

components corresponding to the coefficients are shown in Figure 2.13. To determine an appropriate cutoff level $j = J$, we observe that noise can be described by

$$d_k^{(j)} \sim (-1)^k ,$$

and that $\sum_k (-1)^k \, \psi(t-k)$ is periodic, with period $T = 2$. $g_j(t)$ represents a component with frequency $\omega \sim 2^j \pi$. We use Figure 2.12 to estimate the width of the peak in the acceleration data. $(T/2) \sim 30$ or $\omega = i(2\pi/T) \sim (\pi/30)$, which corresponds to the decomposition level $j \sim -5$. We set the cutoff level to be $J = -4$ and discard $g_j(t)$; $j \leq J = -4$, which we regard as noise. The smoothed acceleration is $f_{-4}(t)$, shown in Figure 2.14.

Table 2.1. *Impact time and energy absorption.*

mother function	height (cm)	t_0	energy (J)	error (%)
Spline 5	3	254 ± 3	0.761 ± 0.004	0.5
	6	320 ± 1	1.366 ± 0.001	0.1
	9	309 ± 2	2.148 ± 0.016	0.8
Spline 4	3	253 ± 3	0.760 ± 0.003	0.4
	6	320 ± 3	1.422 ± 0.001	0.1
	9	321 ± 1	2.213 ± 0.006	0.3
Daubechies' 8	3	254 ± 3	0.751 ± 0.004	0.5
	6	317 ± 2	1.393 ± 0.002	0.1
	9	318 ± 2	2.172 ± 0.009	0.4

A close examination of the acceleration $a(t)$ reveals that there are two regions where $a(t)$ is almost a linear function of t: Region I is when t is less than $\simeq 255$, and Region II is when t is between $\simeq 255$ and $\simeq 300$ (see Figure 2.15). In Region III, in which t is greater than $\simeq 300$, the acceleration $a(t)$ changes continuously. The boundary between Regions I and II is the time of collision t_0. We use least squares fitting of a linear function on the data for $a(t)$ to estimate t_0. Note that there are several possible choices for the data points for each region, which lead to errors in the estimation of t_0. To assess the accuracy of our method, we compare results from using quartic B-spline ($m = 5$), cubic spline ($m = 4$), and Daubechies' ($N = 8$) wavelets. Our results are given in Table 2.1.

We compute the velocity $\dot{x}(t)$ using finite differences and find that it decreases rapidly at $t \sim 320$ ($\times$ 0.0001 sec) (see Figure 2.16). These results show that after the weight hits the surface of the silicone gel at time $t_0 \simeq 254$, there is some delay before the velocity is affected. The length of the time delay is a function of the properties of the gel. Only the cleaner data clearly show the existence of a delay and its duration.

We use the time of collision t_0 as the initial time to produce an x-a plot. The origin $(x(t_0), a(t_0))$ corresponds to $(0, -g)$, where g is the gravitational acceleration. Our convention is that $a(t)$ is positive upward so that $a(t_0) = -g$ at the the time of collision. Figure 2.17 illustrates the effectiveness of our wavelet-based noise reduction method.

We can use the smooth curve to compute the energy absorbed by the gel by computing the area of the triangle determined by two successive points and the origin, which is the determinant

$$\Delta S_n = \frac{1}{2} \begin{vmatrix} 0 & 0 & 1 \\ x(2^j n) & a(2^j n) & 1 \\ x(2^j(n+1)) & a(2^j(n+1)) & 1 \end{vmatrix}.$$

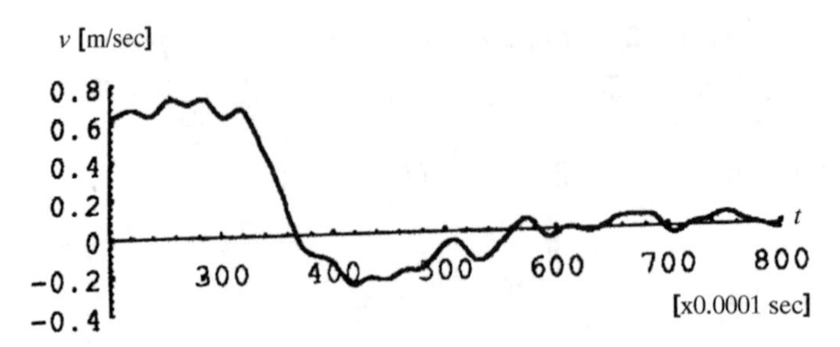

Figure 2.16. *Velocity curve.*

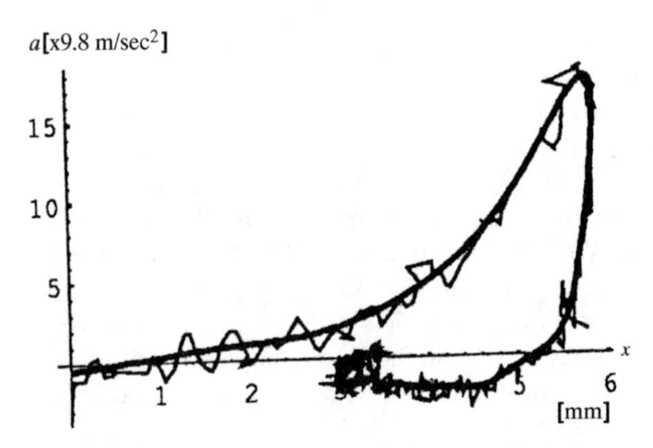

Figure 2.17. *Smooth curve in the xf-plane.*

The energy absorbed by the gel is $m\sum_n \Delta S_n$. Our results from using the points with $j = -4$ are given in Table 2.1. The errors in t_0, as well as the energy absorption, are very small. We note that further reduction in noise and error can be achieved by refining our experimental apparatus and by using an additional sensor to identify, more precisely, the time of collision of the weight and the surface of the silicone gel.

2.4. Conclusion

In this chapter we present a simple and effective wavelet-based method for reducing noise in laboratory data which represent the displacement of a mechanical part. Our method assumes that data are reasonably smooth functions of time and that high level components from a wavelet decomposition method proposed by Mallat consist primarily of noise. It follows that noise can be reduced by subtracting these high level components from the original data. The cutoff threshold is determined through scientific experience and knowledge of properties of clean data.

We conducted two laboratory experiments to show how our wavelet-based noise reduction algorithm can reduce noise to yield smooth functions which are useful for studying properties of mechanical systems.

References

[1] A. Aldroubi, M. Unser (eds.) (1996), *Wavelets in Medicine and Biology*, CRC Press, Tokyo.

[2] A. Antoniadis, G. Oppenheim (eds.) (1995), *Wavelets and Statistics*, Springer-Verlag, Tokyo.

[3] B. Barsky (1988), *Computer Graphics and Geometric Modeling Using Beta-splines*, Springer-Verlag, Tokyo.

[4] B. Bartles, J. Beatty, B. Barsky (1987), *An Introduction to Splines for Use in Computer Graphics and Geometric Modeling*, Morgan-Kaufmann, San Francisco, CA.

[5] C. Chui (1992), *An Introduction to Wavelets*, Academic Press, Tokyo.

[6] R. Coifman, D. Donoho (1995), "Translation invariant de-noising," pp. 125–150 in *Wavelets and Statistics*, Springer-Verlag, Tokyo.

[7] I. Daubechies (1992), *Ten Lectures on Wavelets*, SIAM, Philadelphia, PA.

[8] B. Delyon, A. Juditsky (1995), "Estimating wavelet coefficients," pp. 151–168 in *Wavelets and Statistics*, Springer-Verlag, Tokyo.

[9] D. Donoho, I. Johnston (1994), "Ideal spatial adaptation via wavelet shrinkage," *Biometrica*, vol. 81, pp. 425–455.

[10] G. Farin (1990), *Curves and Surfaces for Computer Aided Geometric Design*, Academic Press, Tokyo.

[11] M. Hilton et al. (1996), "Wavelet denoising of functional MRI data," pp. 93–114 in *Wavelets in Medicine and Biology*, A. Aldroubi, M. Unser (eds.), CRC Press, Tokyo.

[12] M. Malfait (1996), "Using wavelets to suppress noise in biomedical images," pp. 192–209 in *Wavelets in Medicine and Biology*, A. Aldroubi, M. Unser (eds.), CRC Press, Tokyo.

[13] S. Mallat (1989), "Multifrequency channel decompositions of images and wavelet methods," *IEEE Trans. Acoust. Speech Signal Process.*, vol. 37, pp. 2091–2110.

[14] N. Saito (1994), "Simultaneous noise suppression and signal compression using a library of orthonormal bases and the minimum description length criterion," pp. 299–324 in *Wavelets in Geophysics*, E. Georgiou, P. Kumar (eds.), Academic Press, Tokyo.

[15] S. Sakakibara (1994), "A practice of data smoothing by B-spline wavelets," pp. 179–196 in *Wavelets: Theory, Algorithms, and Applications*, C. Chui et al. (eds.), Academic Press, Tokyo.

[16] S. Sakakibara, N. Shimizu (1994), "Wavelet analysis of dry friction force," *Proc. Japanese Soc. Mech. Engineers Mech. Measurement Control*, (A) 940-26, pp. 172–175.

[17] S. Sakakibara, T. Endo (1995), "An analysis of drop mass test using spline wavelets," *Proc. Japanese Soc. Mech. Engineers Mech. Measurement Control*, (B) 95-8(I), pp. 62–65.

[18] E. Shikin, A. Plis (1995), *Handbook on Splines for the User*, CRC Press, Tokyo.

3. A Wavelet-Based Conjugate Gradient Method for Solving Poisson Equations

Nobuatsu Tanaka*

Abstract. This chapter describes a powerful and simple new wavelet-based preconditioning method for solving large systems of linear equations and shows how it is implemented in simulations of fluid flow modeled by Poisson equations. The linear systems can be solved with an iterative matrix solver; however, preconditioning using wavelet bases prevents a marked increase in computing time with respect to an increase in grid points. Use of our technique leads to a matrix with a bounded condition number so that computing time is reduced significantly. Our technique also overcomes difficulties associated with earlier wavelet-based solvers, which assume periodic boundary conditions, making them difficult to apply to real simulation problems. Results from our numerical experiments with one- and two-dimensional Poisson equations confirm the power and accuracy of our method. Unlike many preconditioning methods which are not suitable for vector and parallel processing, our algorithm can take advantage of the extra processing capabilities and enhance computing performance. For example, a speedup of over 100 fold can be achieved when solving Poisson equations on a Cray T3D using 128 processors in parallel.

Key words. wavelet, preconditioning, Poisson equation, matrix solver, conjugate gradient method

3.1. Introduction

Numerical simulations play an important role in the analysis of many kinds of complex phenomena which occur in nuclear power plants, such as neutron transport, fluid flow, and structural behavior. The need for and use of detailed and accurate numerical simulations have been increasing in many engineering fields since the cost of performing physical experiments is considerably higher. This trend is also fueled by the development of more powerful computer hardware, such as vector supercomputers and massively

*Nuclear Engineering Laboratory, Toshiba Corporation, 4-1, Ukishima-cho, Kawasaki-ku, Kawasaki, Kanagawa 210-0862 Japan (nobuatsu.tanaka@toshiba.co.jp).

parallel processors. However, physical experiments cannot be eliminated altogether; we must continue to conduct them to some extent to understand physical phenomena, to assist in drawing a correlation between simulation models and computer codes, and to obtain data for the qualification of these codes.

Scientists need simple mechanistic models with few assumptions that yield reliable and accurate results in numerical simulations. Sophisticated models are often inappropriate or impossible in the fluid flow analysis of nuclear power plants. For example, when the Direct Numerical Simulation Method (one of the most detailed models of fluid flow) is used to simulate nuclear power plant conditions with high Reynolds number, turbulent flow, and systems with complex geometries, a very fine mesh is needed, and computing time increases drastically with an increase in mesh points. The most time-consuming step of the calculation is solving continuous equations governing phenomena. To solve these equations, scientists first approximate them by discretization then transform them into a large system of linear equations. Although the linear system can be solved using a suitable iterative matrix solver, the convergence speed deteriorates with an increase in the coefficient matrix condition number $\kappa(\boldsymbol{A})$, where

$$\kappa(\boldsymbol{A}) = \|\boldsymbol{A}\| \cdot \|\boldsymbol{A}^{-1}\|$$

for a specified matrix norm $\| \cdot \|$. The condition number increases markedly with an increase in the number of mesh points. Recommended references on large matrix computations are [8], [12], and [22].

Two approaches can be used to repress the increase in computing time: the computational algorithm can be improved (e.g., by reducing the number of iterations) or more powerful hardware can be employed (e.g., powerful vector or parallel supercomputers). We use both approaches in our study to obtain significantly better results than have been found in previous works; a new algorithm is developed to reduce computational costs, and the algorithm is implemented on a high-performance computer. Our algorithm suppresses the increase of the computing time by applying a wavelet-based preconditioner to the coefficient matrix. This approach follows from an observation that the condition number can be bounded within a limited value range by using wavelet-based techniques [2], [11].

Wavelets are an important new mathematical tool for describing complex functions and analyzing empirical continuous data derived from many different types of signals. In 1988 Daubechies proposed families of compactly supported wavelets with user-specifiable degrees of smoothness, constructed from a multiresolution analysis (MRA) [4]. Studies by researchers using these wavelets and others for solving partial differential equations include

[2], [3], [9], and [11]. In one study, Beylkin proposes a new algorithm to solve a linear system using the discrete wavelet transform (DWT) which prevents an increase in the condition number of a large matrix [2]. We modify his method—a method which is difficult to implement to solve practical problems, because it assumes periodic boundary conditions (BCs). Our technique, the incomplete discrete wavelet transform (iDWT), does not assume periodic BCs [17], [19]. Instead, it assumes that all values outside the boundaries are equal to zero. Our method suppresses the increase in the condition number of a large matrix through diagonal rescaling. The effects and advantages of the iDWT preconditioning are confirmed with one-dimensional boundary value problems (BVPs) of elliptic equations. Extensions to multidimensional problems are also investigated.

The remainder of this chapter is organized as follows. In the next section, we begin with a brief review of MRAs and wavelets, then one- and two-dimensional iDWT algorithms are given. The third section describes our implementation studies using the Cray C94D vector and Cray T3D parallel supercomputers. The fourth and concluding section is a summary of our findings.

3.2. The Incomplete Discrete Wavelet Transform

In this section we review basic concepts associated with scaling functions and wavelets. MRA and basis sets for subspaces of $L^2(\boldsymbol{R})$ are described. We select scaling functions and wavelets to be the basis sets and review how they can be used to approximate functions in $L^2(\boldsymbol{R})$. The main portion of this section consists of a presentation of our one- and two-dimensional incomplete discrete wavelet transform (iDWT) algorithm.

3.2.1. One-Dimensional Problem

Consider the one-dimensional elliptic equation

$$\begin{aligned} -\Delta u &= f \quad \text{in} \quad \Omega = [0,1], \\ u &= g \quad \text{on} \quad \partial\Omega, \\ \frac{\partial u}{\partial x} &= 0 \quad \text{on} \quad \partial\Omega. \end{aligned} \tag{3.1}$$

We discretize this problem using a uniform mesh to obtain a system of linear equations of the form

$$\boldsymbol{A}\boldsymbol{u} = \boldsymbol{f}, \tag{3.2}$$

where $\boldsymbol{u} = \{u_i\}, \boldsymbol{f} = \Delta x^2\{f_i\}$, the coefficient matrix $\boldsymbol{A}$ is a 2^n-by-2^n, tridiagonal, positive-definite matrix of the form

$$\boldsymbol{A} = \underbrace{\left.\begin{pmatrix} a & -1 & 0 & \cdots & 0 \\ -1 & 2 & -1 & & \vdots \\ 0 & -1 & 2 & \ddots & 0 \\ \vdots & & \ddots & \ddots & -1 \\ 0 & \cdots & 0 & -1 & b \end{pmatrix}\right\}}_{2^n} 2^n,$$

and a and b are real numbers which represent the BCs at each edge point. Hereafter, we assume $N = 2^n$.

Before describing the iDWT, we discuss how an MRA can be used to approximate a function via a scaling function and wavelet expansion. (The definition of an MRA is given in Chapter 1, section 1.2 of this book.) Let $\{V_j : j \in \boldsymbol{Z}\}$ in $L^2(\boldsymbol{R})$ be an MRA. Then there exists an orthonormal wavelet basis $\{\psi_{j,k} : j, k \in \boldsymbol{Z}\}$ of $L^2(\boldsymbol{R})$ such that

$$\psi_{j,k}(x) = 2^{-j/2}\ \psi(2^{-j}x - k), \tag{3.3}$$

and

$$P_{j-1}f = P_j f\ +\ \sum_{k\in\boldsymbol{Z}} \langle f, \psi_{j,k}\rangle\ \psi_{j,k},$$

for any f in $L^2(\boldsymbol{R})$, where P_j is the orthogonal projection onto V_j [5]. For every $j \in \boldsymbol{Z}$, let W_j be the orthogonal complement of V_j in V_{j-1}, then $V_{j-1} = V_j \oplus W_j$ and $W_j \perp W_{j'}$ for $j \neq j'$ so that

$$V_j = V_J \oplus \bigoplus_{k=0}^{J-j-1} W_{J-k} \quad \text{for} \quad j < J,$$

$$L^2(\boldsymbol{R}) = \bigoplus_{j\in\boldsymbol{Z}} W_j.$$

In this study, we use the basis

$$\psi_{j,k}(x) = 2^{(n-j)/2}\ \psi(2^{n-j}x - k)$$

$$
\begin{array}{ccccccccc}
\{s_k^0\} & \longrightarrow & \{s_k^1\} & \longrightarrow & \{s_k^2\} & \longrightarrow & \cdots & \longrightarrow & \{s_k^J\} \\
& \searrow & & \searrow & & \searrow & & \searrow & \\
& & \{d_k^1\} & & \{d_k^2\} & & \cdots & & \{d_k^J\}
\end{array}
$$

Figure 3.1. *Pyramid algorithm.*

at the basic resolution level $j = 0$, instead of using equation (3.3). $\varphi_{j,k}$ and $\psi_{j,k}$ are orthonormal bases of V_j and W_j, respectively, where

$$
\begin{aligned}
\varphi_{j,k}(x) &= \sum_i h_k \ \varphi_{j-1,2k+i}(x), \\
\psi_{j,k}(x) &= \sum_i g_k \ \varphi_{j-1,2k+i}(x),
\end{aligned}
$$

and $\{h_i\}$, $\{g_i\}$ are the wavelet filter coefficients, which satisfy $g_i = (-1)^i h_{1-i}$. Any function $u(x) \in L^2(\Omega)$ can be approximated by a scaling function and wavelet expansion as

$$
u(x) \cong \sum_k s_k^J \ \varphi_{J,k}(x) \ + \ \sum_{j=1}^{J} \sum_k d_k^j \ \psi_{j,k}(x), \tag{3.4}
$$

where s_k^j and d_k^j are the coefficients of the expansion, and

$$
s_k^j = \int_\Omega u(x) \ \varphi_{j,k}(x) \, dx, \quad d_k^j = \int_\Omega u(x) \ \psi_{j,k}(x) \, dx.
$$

The coefficients of basic resolution level ($j = 0$) are

$$
s_k^0 = 2^{n/2} \int_\Omega u(x) \ \varphi(2^n x - k) \, dx.
$$

The support of $\varphi(x)$ is the interval $[0, 2M-1]$, and $\varphi(0) = \varphi(2M-1) = 0$, so the integral above can be approximated by the sum

$$
s_k^0 \cong 2^{-n/2} \sum_{i=1}^{2M-2} u_{k+i} \varphi(i). \tag{3.5}
$$

When n is sufficiently large and $u(x)$ sufficiently smooth, we use an even simpler approximation

$$
s_k^0 \cong 2^{-n/2} \ u_k
$$

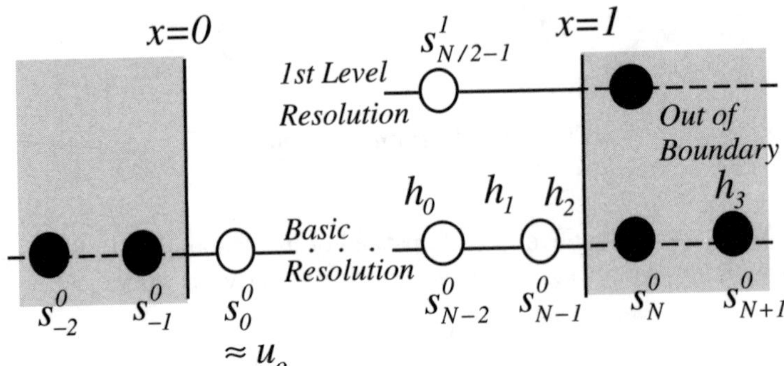

Figure 3.2. *The pyramid algorithm and boundary points.*

for the sum. This simpler expression allows us to generate the coefficients $\{s_k^0\}$ using only the data $\{u_i\}$ and $\{f_i\}$ $(\times 2^{-n/2})$ which are defined on the grid nodes. The coefficients in equation (3.4) are evaluated using the pyramid algorithm (illustrated in Figure 3.1) given by the equations

$$s_k^j = \sum_i h_i \, s_{2k+i}^{j-1}, \quad d_k^j = \sum_i g_i \, s_{2k+i}^{j-1}.$$

Straightforward implementation of the pyramid algorithm leads to difficulties in handling boundary points, since the procedure requires several data values which are defined outside the boundaries and are not known (as shown in Figure 3.2). To overcome this difficulty, Beylkin [2] and Koornwinder [14] introduced a periodic BC requirement. Since the periodized wavelet-based transformations are orthonormal, we refer to their method as the complete discrete wavelet transform (cDWT). However, the requirement introduces a new difficulty, namely, a means for transforming general BVPs to problems with periodic BCs. Short of a simple answer, the method is too complicated to adopt for solving general problems.

Our iDWT method overcomes the problems associated with the cDWT method, because it does not require periodic BCs and assumes that all values outside the boundaries are equal to zero. Although the transformations are not orthonormal, the algorithm is more suitable and effective than its complete counterpart for numerical simulations and is much simpler to implement.

Let $\boldsymbol{W}_{(J)}$ denote the iDWT matrix which transforms coefficient data from the basic to the Jth order resolution. Then

$$\boldsymbol{W}_{(J)} = \boldsymbol{W}_{J-1} \, \boldsymbol{W}_{J-2} \, \cdots \, \boldsymbol{W}_0, \tag{3.6}$$

where $\boldsymbol{W}_j$ is the $N \times N$ matrix which transforms data from the jth to the $(j+1)$th resolution, i.e.,

$$\boldsymbol{W}_j = \begin{pmatrix} \begin{pmatrix} \boldsymbol{H}_j \\ \boldsymbol{G}_j \end{pmatrix} & \boldsymbol{0} \\ \boldsymbol{0} & \boldsymbol{I}_{N-2^{n-j}} \end{pmatrix},$$

$\boldsymbol{I}_i$ is the $i \times i$ unit matrix, and $\boldsymbol{H}_j$ and $\boldsymbol{G}_j$ are 2^{n-j}-by-2^{n-j-1} banded matrices without the periodized elements of the cDWT (lower left elements of $\boldsymbol{H}_j$ and upper right elements of $\boldsymbol{G}_j$):

$$H_j = \underbrace{\left.\begin{pmatrix} h_0 & h_1 & h_2 & \cdots & h_{2M-1} & 0 & \cdots & & 0 & 0 \\ 0 & 0 & h_0 & h_1 & h_2 & \cdots & h_{2M-1} & 0 & & 0 \\ \vdots & & & \ddots & & & & \ddots & & \vdots \\ 0 & & \cdots & 0 & h_0 & h_1 & & \cdots & & h_{2M-1} \\ \vdots & & & & & \ddots & & & & \vdots \\ 0 & & & & 0 & 0 & h_0 & h_1 & h_2 & h_3 \\ 0 & 0 & & & \cdots & & 0 & 0 & h_0 & h_1 \end{pmatrix}\right\} 2^{n-j-1}}_{2^{n-j}}$$

and

$$\boldsymbol{G}_j = \underbrace{\left.\begin{pmatrix} g_0 & g_1 & 0 & 0 & 0 & & \cdots & & 0 & 0 \\ g_{-2} & g_{-1} & g_0 & g_1 & 0 & 0 & & & & 0 \\ \vdots & & & & \ddots & & & & & \vdots \\ g_{-2M} & & \cdots & & g_0 & g_1 & 0 & \cdots & & 0 \\ 0 & \ddots & & & & & \ddots & & & \vdots \\ \vdots & & 0 & g_{-2M} & \cdots & g_{-1} & g_0 & g_1 & 0 & 0 \\ 0 & 0 & \cdots & 0 & 0 & g_{-2M} & \cdots & g_{-1} & g_0 & g_1 \end{pmatrix}\right\} 2^{n-j-1}}_{2^{n-j}}.$$

Equation (3.2) can be expressed in terms of these matrices as

$$\boldsymbol{W}_{(J)}\, \boldsymbol{A}\, \boldsymbol{W}_{(J)}^{-1}\, \vec{\boldsymbol{u}}_J = \vec{\boldsymbol{f}}_J,$$

where

$$\vec{\boldsymbol{u}}_J = 2^{-n/2}\boldsymbol{W}_{(J)}\, \boldsymbol{u}, \qquad \vec{\boldsymbol{f}}_J = 2^{-n/2}\boldsymbol{W}_{(J)}\, \boldsymbol{f}.$$

Both $\vec{u}_J$ and $\vec{f}_J$ are composed of wavelet expansion coefficients of $u(x)$ and $f(x)$. Since $W_{(J)}$ approximates the cDWT, which is an orthonormal transform, i.e.,

$$W_{(J)}\, W_{(J)}^T \cong I_N,$$

so that

$$W_{(J)}\, A\, W_{(J)}^T\, \vec{u}_J \cong \vec{f}_J. \tag{3.7}$$

3.2.2. Diagonal Rescaling Method

When the linear system (3.2) is transformed into a space with wavelets as basis functions, the matrix A which represents the Laplacian operator is transformed into a matrix $W_{(J)}\, A\, W_{(J)}^T$. Since the iDWT is approximately orthonormal, the condition number of a transformed matrix does not improve. However, diagonal rescaling is very effective in reducing the condition number of the transformed matrix. The eigenfunctions and eigenvalues for the Laplacian operator Δ in W_j are

$$E_\xi^j = \sum_k e^{ik\xi}\, \psi_k^j$$

and

$$\lambda_\xi^j = 2^{2(n-j)} \sum_k q_k\, e^{-ik\xi},$$

respectively, where

$$q_k = \int_\Omega \psi\,(x-k)\, \frac{d^2}{dx^2}\, \psi\,(x)\, dx,$$

$\Omega = \boldsymbol{R}$, $0 \le \xi < 2\pi$, and the summation with respect to the index k is over all integers [9]. When the functions are periodic in Ω, ξ assumes the discrete values $\{(2\pi k)/2^n : k = 0, 1, 2, \dots, 2^n - 1\}$, and the diagonal rescaling matrix takes the form

$$P_j = \begin{pmatrix} 2^j I_{2^{n-j+1}} & 0 & 0 & \cdots & 0 \\ 0 & 2^{j-1} I_{2^{n-j+1}} & 0 & & 0 \\ 0 & 0 & 2^{j-2} I_{2^{n-j+2}} & & \vdots \\ \vdots & & & \ddots & 0 \\ 0 & 0 & \cdots & 0 & 2 I_{2^{n-1}} \end{pmatrix}$$

for $j > 1$, and

$$\boldsymbol{P}_j = 2\,\boldsymbol{I}_{2^n}$$

for $j = 1$. We apply the matrix $\boldsymbol{P}_J$ to system (3.7) to obtain

$$\boldsymbol{P}_J\,\boldsymbol{W}_{(J)}\,\boldsymbol{A}\,\boldsymbol{W}_{(J)}^T\,\boldsymbol{P}_J\bar{\boldsymbol{u}}_J \cong \bar{\boldsymbol{f}}_J, \tag{3.8}$$

where

$$\bar{\boldsymbol{u}}_J = 2^{-n/2}\boldsymbol{P}_J^{-1}\,\boldsymbol{W}_{(J)}\,\boldsymbol{u}, \qquad \bar{\boldsymbol{f}}_J = 2^{-n/2}\boldsymbol{P}_J\,\boldsymbol{W}_{(J)}\,\boldsymbol{f}.$$

System (3.8), which has been diagonally rescaled, is now numerically applicable to the preconditioned conjugate gradient (PCG) method (described in Appendix 2). Since rescaling improves the condition number, the increase in the number of PCG iterations associated with the increase in grid points is suppressed.

3.2.3. Incomplete discrete wavelet transform pre-conditioned conjugate gradient (IDWT PCG)

We apply the iDWT to the PCG method by transforming (3.2) into the form

$$\boldsymbol{P}_J\,\boldsymbol{W}_{(J)}\,\boldsymbol{A}\,\boldsymbol{W}_{(J)}^T\,\boldsymbol{P}_J\,\hat{\boldsymbol{u}}_J = \hat{\boldsymbol{f}}_J, \tag{3.9}$$

where

$$\hat{\boldsymbol{u}}_J = 2^{-n/2}\boldsymbol{P}_J^{-1}\boldsymbol{W}_{(J)}^{-T}\boldsymbol{u}, \qquad \hat{\boldsymbol{f}}_J = 2^{-n/2}\,\boldsymbol{P}_J\,\boldsymbol{W}_{(J)}\,\boldsymbol{f}.$$

Note that while equation (3.8) approximates (3.2), equation (3.9) is equivalent to (3.2). To solve (3.9) using the PCG method, set $\boldsymbol{V} = \boldsymbol{P}_j\boldsymbol{W}_{(j)}$ and the preconditioning matrix $\boldsymbol{K}$ to

$$\boldsymbol{K} = \boldsymbol{V}^T\,\boldsymbol{V} = \boldsymbol{W}_{(J)}^T\,\boldsymbol{P}_J^2\,\boldsymbol{W}_{(J)}.$$

For a one-dimensional Poisson equation solver, the operation $\boldsymbol{P}_j^2$ can be simplified to the change in filter coefficients

$$\{h_k'\} = 2\,\{h_k\} \quad \text{and} \quad \{g_k'\} = 2\,\{g_k\}.$$

We compute the preconditioned residual,

$$\vec{\boldsymbol{r}} = \boldsymbol{W}_{(J)}^T\,\boldsymbol{P}_J^2\,\boldsymbol{W}_{(J)}\,\boldsymbol{r},$$

for the PCG algorithm using this change in filter coefficients. The system is then solved using the PCG method.

3.2.4. Two-Dimensional Problem

Consider the two-dimensional Poisson equation,

$$\begin{aligned} -\Delta u &= f \quad \text{in} \quad \Omega = [0,1] \times [0,1], \\ u &= g \quad \text{on} \quad \partial\Omega. \end{aligned} \tag{3.10}$$

To solve this problem numerically, we use a finite difference scheme with a uniform mesh on region Ω and the direct product of two, one-dimensional, orthonormal bases to extend the iDWT to two dimensions, i.e.,

$$V_{j-1}^{(xy)} = \left(V_j^{(x)} \otimes V_j^{(y)}\right) \oplus \left(V_j^{(x)} \otimes W_j^{(y)}\right) \oplus \left(W_j^{(x)} \otimes V_j^{(y)}\right) \oplus \left(W_j^{(x)} \otimes W_j^{(y)}\right).$$

Then, the scaling function and wavelet expansion of $u(x,y) \in L^2(\Omega)$ is expanded to

$$\begin{aligned} u(x,y) &\cong \sum_{l,m} s_{l,m}^J \varphi_l^J \varphi_m^J \\ &+ \sum_{j=1}^{J} \sum_{l,m} \left\{ d_{l,m}^{(x),j} \ \psi_l^j \ \varphi_m^j \ + \ d_{l,m}^{(y),j} \ \varphi_l^j \ \psi_m^j \ + \ d_{l,m}^{(xy),j} \ \psi_l^j \ \psi_m^j \right\}, \end{aligned}$$

with expansion coefficients

$$\begin{aligned} s_{l,m}^j &= \sum_{i,k} h_i \ h_k \ s_{2l+i,2m+k}^{j-1}, \\ d_{l,m}^{(x),j} &= \sum_{i,k} g_i \ h_k \ s_{2l+i,2m+k}^{j-1}, \\ d_{l,m}^{(y),j} &= \sum_{i,k} h_i \ g_k \ s_{2l+i,2m+k}^{j-1}, \\ d_{l,m}^{(xy),j} &= \sum_{i,k} g_i \ g_k \ s_{2l+i,2m+k}^{j-1}. \end{aligned}$$

The eigenvalues and eigenfunctions of the two-dimensional Laplacian operator $\Delta \equiv (\partial^2/\partial x^2 + \partial^2/\partial x^2)$ are

$$E_{\xi,\eta}^j = \sum_{l,m} e^{i(l\xi+m\eta)} \ \varphi_l^j \ \psi_m^j$$

and

$$\lambda^j_{\xi,\eta} = 2^{2(n-j)} \sum_k \left(p_k \, e^{-ik\xi} \; + \; q_k \, e^{-ik\eta} \right),$$

respectively, where

$$p_k = \int_\Omega \varphi\,(x-k)\; \frac{d^2}{dx^2}\,\varphi\,(x)\; dx.$$

As in the one-dimensional case, diagonal rescaling for the two-dimensional Poisson equation amounts to transformation of filter coefficients in each coordinate

$$\{h_i'\} = \sqrt{2}\,\{h_i\} \quad \text{and} \quad \{g_i'\} = \sqrt{2}\,\{g_i\}\,.$$

3.3. Numerical Examples

Our numerical implementations of the iDWT algorithm for one- and two-dimensional problems are described in this section. We compare our results with those obtained from using standard techniques, such as the conjugate gradient (CG), successive over-relaxation (SOR), incomplete Cholesky conjugate gradient (ICCG), and Beylkin's wavelet-based methods. Implementation studies using Cray C94D vector and Cray T3D parallel supercomputers are also described.

3.3.1. One-Dimensional Problem

We consider the numerical solution of the one-dimensional Poisson equation. Table 3.1 shows results from experiments by Beylkin [2] in which a uniform mesh of N grid points were used to solve the periodized BVP, with and without cDWT preconditioning, using Daubechies' orthonormal wavelets with three vanishing moments [4]. Table 3.2 shows our experimental results for the Dirichlet BVP using iDWT preconditioning and the same wavelets.

The results show that the iDWT preconditioning method effectively suppresses the increase in the condition number when the number of grid points is increased. To determine the underlying reason, we analyze the distribution of the eigenvalues for the original and iDWT preconditioned matrices for computations using 256 grid points (see Figure 3.3). Note that the eigenvalues of the original matrix are uniformly distributed. iDWT preconditioning shifts most of the eigenvalues closer together, towards the median

Table 3.1. $\kappa(\boldsymbol{A})$ *for the periodized BVP with and without cDWT preconditioning* ($M = 3,\ J = n$), *from Beylkin* [2].

grid points	$\boldsymbol{A}$	$\boldsymbol{P}_J \boldsymbol{W}_{(J)} \boldsymbol{A} \boldsymbol{W}_{(J)}^T \boldsymbol{P}_J$
32	104	8.021
64	415	9.086
128	1660	10.02
256	6640	10.84

Table 3.2. $\kappa(\boldsymbol{A})$ *for the Dirichlet BVP with and without iDWT preconditioning* ($a = b = 2,\ M = 3,\ J = n - 2$).

grid points	$\boldsymbol{A}$	$\boldsymbol{P}_J \boldsymbol{W}_{(J)} \boldsymbol{A} \boldsymbol{W}_{(J)}^T \boldsymbol{P}_J$
32	440.7	27.0
64	1713	53.7
128	6742	107.6
256	26800	216.0

range 10–100. This new distribution prevents an increase in the condition number as the number of grid points increases and reduces the number of iterations and computing time required for convergence of conjugate gradient (CG) algorithms.

Next, we investigate the effect of the iDWT on the condition numbers of matrices associated with the Dirichlet, mixed, and Neumann BVPs. Mixed BVPs are those with one boundary satisfying Dirichlet conditions and another boundary satisfying Neumann conditions. Our results using Daubechies' wavelets with three vanishing moments are shown in Tables 3.3 and 3.4. In a Neumann BVP, the second smallest eigenvalue rather than the smallest is used to compute the condition number, because the smallest is zero. In both BVPs, preconditioning suppresses the increase in the condition number when the number of grid points is increased.

To verify that the iDWTCG method subdues the increase in the number of iterations required for convergence and computing time, we conducted experiments for the Dirichlet problem with the convergence criterion

$$\|\boldsymbol{A}\boldsymbol{u}_i - \boldsymbol{f}\|_2 \ \le\ 1.0 \times 10^{-8} \|\boldsymbol{f}\|_2$$

(see Table 3.5). The number of iterations required for convergence of the CG method without preconditioning increases in direct proportion to the increase in grid points, whereas, for the iDWTCG method, the increase is very slight.

Table 3.3. $\kappa(\boldsymbol{A})$ *for the mixed BVP (Dirichlet and Neumann)* $(a = 2,\ b = 1,\ M = 3,\ J = n - 2)$.

grid points	$\boldsymbol{A}$	$\boldsymbol{P}_J \boldsymbol{W}_{(J)} \boldsymbol{A} \boldsymbol{W}_{(J)}^T \boldsymbol{P}_J$
32	1709	33.96
64	6743	40.90
128	26700	47.36
256	107000	53.20

Table 3.4. $\kappa(\boldsymbol{A})$ *for the Neumann BVP* $(a = b = 1,\ M = 3,\ J = n - 2)$.

grid points	$\boldsymbol{A}$	$\boldsymbol{P}_J \boldsymbol{W}_{(J)} \boldsymbol{A} \boldsymbol{W}_{(J)}^T \boldsymbol{P}_J$
32	414.4	28.90
64	1659	37.59
128	6645	45.32
256	26480	51.99

Table 3.5. *Iterations and CPU required to solve the Dirichlet BVP* $(a = b = 2,\ M = 3,\ j = n - 2)$.

grid points	iterations		CPU time (sec.)	
	CG	iDWTCG	CG	iDWTCG
256	253	41	0.048	0.055
512	507	48	0.19	0.12
1024	1014	53	0.95	0.27
2048	2030	58	5.3	0.74
4096	4065	63	23.7	1.80

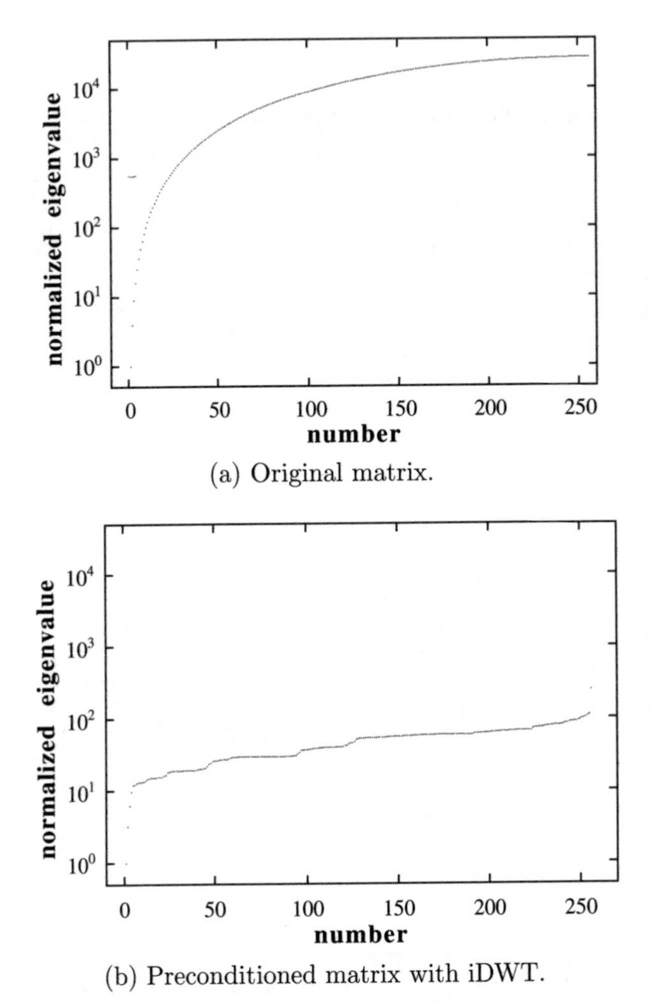

(a) Original matrix.

(b) Preconditioned matrix with iDWT.

Figure 3.3. *Distribution of normalized eigenvalues.*

3.3.2. Two-Dimensional Problem

3.3.2.1. Basic Characteristics of Major Iterative Solvers

Our studies described above show the effectiveness of iDWT preconditioning for solving the one-dimensional Poisson equation. We now consider the two-dimensional problem, since two or more dimensions are normally required for modeling real fluid dynamics problems. We begin by analyzing the theoretical performance of two basic solvers, the CG and SOR methods for solving problem (3.11). (The basics of CG and SOR methods are given in Appendices 1 and 3, respectively.) The BCs are assumed to be Dirichlet, so the eigenvalues of the matrix $\boldsymbol{A}$ which represent the discretized Laplacian operator are

$$\lambda = \frac{1}{2}\,(2 - \cos i\pi\Delta h - \cos j\pi\Delta h)\,; \qquad i, j = 1, 2, \ldots, N.$$

First we consider the CG method. The condition number κ for the matrix $\boldsymbol{A}$ from (3.11) is

$$\kappa(\boldsymbol{A}) = \frac{\lambda_{\max}}{\lambda_{\min}} = \frac{1 + \cos\pi\Delta h}{1 - \cos\pi\Delta h}\,.$$

The error in the approximate solution determined by the CG method after the kth step satisfies

$$\|\boldsymbol{u} - \boldsymbol{u}_k\|_{\boldsymbol{A}} \leq \left(\frac{\sqrt{\kappa - 1}}{\sqrt{\kappa + 1}}\right)^k \|\boldsymbol{u} - \boldsymbol{u}_0\|_{\boldsymbol{A}}\,,$$

where the energy norm $\|\cdot\|_{\boldsymbol{A}}$, defined by

$$\|\boldsymbol{x}\|_{\boldsymbol{A}} = \|\boldsymbol{A}^{1/2}\,\boldsymbol{x}\|_2 = \langle \boldsymbol{A}\boldsymbol{x}, \boldsymbol{x}\rangle^{1/2},$$

is used to measure the residual [6]. If we assume the convergence criterion ε such that

$$\|\boldsymbol{u} - \boldsymbol{u}_k\|_{\boldsymbol{A}} \;\leq\; \varepsilon\|\boldsymbol{u} - \boldsymbol{u}_0\|_{\boldsymbol{A}},$$

then the number of iterations required for convergence of the CG method without preconditioning is proportional to N, i.e.,

$$k \;\leq\; -\,\frac{\log\varepsilon}{\pi}\;(N+1)\,.$$

Effective use of SOR solvers depends on selecting the optimal (or near optimal) acceleration parameter for a given problem. The optimal parameter for problem (3.11) is

$$\omega_{opt} = \frac{1}{1 + \sin\pi\Delta h},$$

and the corresponding spectral radius (i.e., the magnitude of the largest eigenvalue of the matrix $\mathbf{A}$) is

$$\rho_{opt} = \frac{1 - \sin\pi\Delta h}{1 + \sin\pi\Delta h}.$$

We use the power series expansion

$$\ln \frac{1-\sin x}{1+\sin x} = -2x - \frac{x^3}{3} - \frac{x^5}{12} - \cdots,$$

for N sufficiently large, to approximate the asymptotic convergence ratio

$$R_\infty = -\ln \rho_{opt} \cong \frac{\pi}{N+1}$$

[6] and conclude that the number of iterations required for convergence is proportional to the number of grid points N, when the optimal acceleration parameter is used. In the general case for a fixed acceleration parameter $\omega < \omega_{opt}$, the corresponding spectral radius is

$$\rho_{const} = \frac{1}{2}\left(2 - 2\omega + \omega^2 \cos^2 \pi\Delta h + \omega \cos \pi\Delta h \sqrt{4 - 4\omega + \omega^2 \cos^2 \pi\Delta h}\right),$$

and the asymptotic convergence ratio

$$R_\infty = -\ln \rho_{const} \cong \frac{\omega}{2-\omega} \cdot \frac{\pi^2}{(N+1)^2}.$$

The number of iterations required for convergence increases proportionately to N^2, i.e., the speed of convergence is worse than the optimal parameter case.

Note that we used different norms for determining the convergence criteria for the CG and SOR methods (the energy and the l^2-norms), but for the purposes of our analysis, which examines convergence of $O(N)$, $O(N^2)$, etc., it does not matter. In fact, there are many types of vector norms. Some of the most commonly used examples for a vector $\boldsymbol{x} = \{x_1, x_2, \dots, x_n\}$ are the Euclidean norm $\|\cdot\|_2$,

$$\|\boldsymbol{x}\|_2 = \langle \boldsymbol{x}, \boldsymbol{x} \rangle^{1/2},$$

the p-norm $\|\cdot\|_p$,

$$\|\boldsymbol{x}\|_p = \left[\sum_i (\boldsymbol{x}_i)^p\right]^{1/p},$$

and the maximum norm $\|\cdot\|_\infty$,

$$\|\boldsymbol{x}\|_\infty = \max_i\{|x_i|\}.$$

Relations exist among the various types of norms, e.g.,

$$\|\boldsymbol{x}\|_\infty \leq \|\boldsymbol{x}\|_2 \leq n \cdot \|\boldsymbol{x}\|_\infty,$$

Table 3.6. *Iterations required to solve the two-dimensional problem.*

N	SOR ($\omega = \omega_{opt}$)	SOR ($\omega = 1.9$)	CG	ICCG (1,1)	iDWTCG ($M = 3, j = n - 2$)
128	543	1643	220	102	59
256	1110	6514	452	175	77
512	2270	25119	931	361	102
1024	not implemented		1910	709	131

where n is a dimension of the vector $\boldsymbol{x}$. In particular,

$$\sqrt{|\lambda|_{\min}}\ \|\boldsymbol{x}\|_2 \ \le\ \|\boldsymbol{x}\|_A = \|\boldsymbol{A}^{1/2}\ \boldsymbol{x}\|_2 \ \le\ \sqrt{|\lambda|_{\max}}\ \|\boldsymbol{x}\|_2.$$

We compare our theoretical results with numerical implementations using

$$\|\boldsymbol{r}_i\| \ = \ \|\boldsymbol{A}\boldsymbol{u}_i - \boldsymbol{f}\|_2 \ \le\ 1.0 \times 10^{-8}\|\boldsymbol{f}\|_2$$

as the convergence criterion. Our results (given in Table 3.6) show that the number of iterations required for the CG and SOR methods using the optimal acceleration parameter are proportional to N, and for SOR with a fixed acceleration, proportional to parameter N^2. They are consistent with our theoretical analysis described above. Furthermore, we observed that the iDWTCG method significantly lowers the rate of increase in the number of iterations required for convergence when the number of grid points increases.

3.3.2.2. Vector Processing

In this section, we present results from our numerical solution of (3.2) using the ICCG(1,1) and iDWTCG methods with Daubechies' wavelets with three vanishing moments on a Cray C94D, with

$$\|\boldsymbol{A}\boldsymbol{u}_i - \boldsymbol{f}\|_2 \ \le\ 1.0 \times 10^{-8}\|\boldsymbol{f}\|_2$$

as the convergence criterion. Our results are illustrated in Figure 3.4. When a small number of grid points is used, the required CPU time for the two methods is almost the same; however, as the number of grid points increases, the iDWTCG method requires far less CPU time. For the relation

$$\text{computing time} \ \propto\ (\text{number of grids in each direction})^p,$$

$p = 2.88$ for the ICCG method and $p = 2.01$ for the iDWTCG method. The CPU time with iDWTCG is approximately proportional to the number of grid points. The results from our experiments confirm that the iDWTCG is an effective tool for solving large-scale problems.

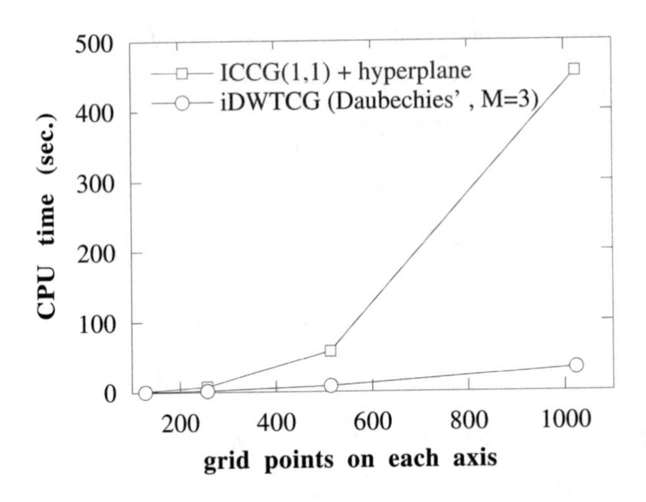

Figure 3.4. *CPU time for ICCG and iDWTCG on Cray C94D.*

3.3.2.3. Parallel Processing

We determined a numerical solution of (3.2) using the iDWTCG on a Cray T3D, a parallel multiple instruction multiple data stream (MIMD) with distributed memory in which processor intercommunication is facilitated by three-dimensional torus networking. The work-sharing model, the simplest of three commonly used types of parallel programming models (the other two being procedure decomposition and data decomposition), was used in our studies.

We parallelized the iDWTCG algorithm by inserting compiler directives. Compiler directives are statements or commands used to change codes to run on vector and parallel machines. For example, for the serial FORTRAN code

```
      do 100 i=1,1000
        x(i) = float(i)**2
      100  continue
```

we insert the compiler directive *cdir* before the DO loop as follows:

```
cdir$ doshared (i) on x(i)
      do 100 i=1,1000
        x(i) = float(i)**2
      100  continue
```

Results from implementations using Daubechies' wavelets with three vanishing moments with 512×512 grid points are shown in Figure 3.5. We

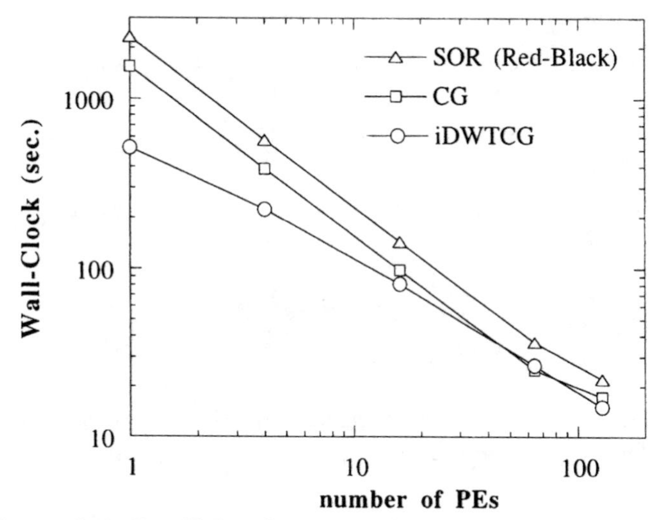

Figure 3.5. *Parallel performance of original iDWTCG solver.*

make it possible for SOR to process in parallel by reducing data recurrence relations with Red-Black ordering.

Red-Black ordering is a technique used during the conversion of serial algorithms to parallel to eliminate inconvenient data dependencies as much as possible. For example, straightforward discretization of the two-dimensional Poisson equation leads to terms such as

$$4\ u_{i,j} - u_{i-1,j} - u_{i+1,j} - u_{i,j-1} - u_{i,j+1} \ = \ h^2\ f_{i,j},$$

for $i, j = 1, 2, 3$, where the $f_{i,j}$ are known, and where the $u_{i,j}$ are unknown. If the SOR method is applied, then

$$u_{i,j} = u_{i,j} - \frac{\omega}{4}\ \left(4\ u_{i,j} - u_{i-1,j} - u_{i+1,j} - u_{i,j-1} - u_{i,j+1} - h^2\ f_{i,j}\right).$$

If we compute $u_{i,j}$ using lexicographical ordering, shown below, the data are related recurrently, i.e., we need to calculate $u_{1,2}$ and $u_{2,1}$ before $u_{2,2}$.

$$\begin{array}{lccccc} & (i=1) & & (i=2) & & (i=3) \\ (j=1) & 1 & - & 2 & - & 3 \\ & | & & | & & | \\ (j=2) & 4 & - & 5 & - & 6 \\ & | & & | & & | \\ (j=3) & 7 & - & 8 & - & 9 \end{array}$$

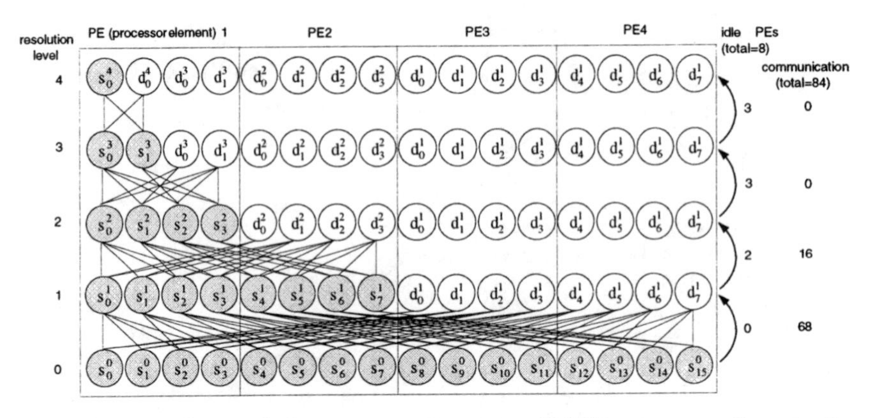

Figure 3.6. *Original data arrangement in iDWT corresponding to the matrix defined by equation* (3.6).

To eliminate inconvenient data interdependencies in a parallel algorithm, the order of computations should be changed. One of the commonly used methods is Red-Black (checkerboard-like) ordering, shown below, in which $i = 1, 2, 3$.

$$
\begin{array}{ccccccc}
 & (i=1) & & (i=2) & & (i=3) \\
(j=1) & 1 & - & 6 & - & 2 \\
 & | & & | & & | \\
(j=2) & 7 & - & 3 & - & 8 \\
 & | & & | & & | \\
(j=3) & 4 & - & 9 & - & 5
\end{array}
$$

Red-Black ordering allows independent, parallel computation of data points $1, 2, 3, 4, 5$ in the Red group and points $6, 7, 8, 9$ in the Black group. Many other types of ordering methods exist in the literature [7].

The required computing times for both the SOR and CG methods decrease at a constant rate with an increase in processor elements (PEs). Although the iDWTCG method solves the problem faster than the others when only one processor is used, its performance improves at a slower rate as additional PEs are used. As a consequence, the required computing time for CG and iDWTCG methods is about the same when 64 PEs are used; preconditioning leads to no improvement in performance.

However, the computing performance of the iDWTCG algorithm, using multiple PEs, improves remarkably when the data arrangement is changed. The DWT uses the data arrangement shown in Figure 3.6 (a one-dimensional case with two vanishing-moments Daubechies' wavelets), which is simple to code. However, this arrangement is not suitable for parallel pro-

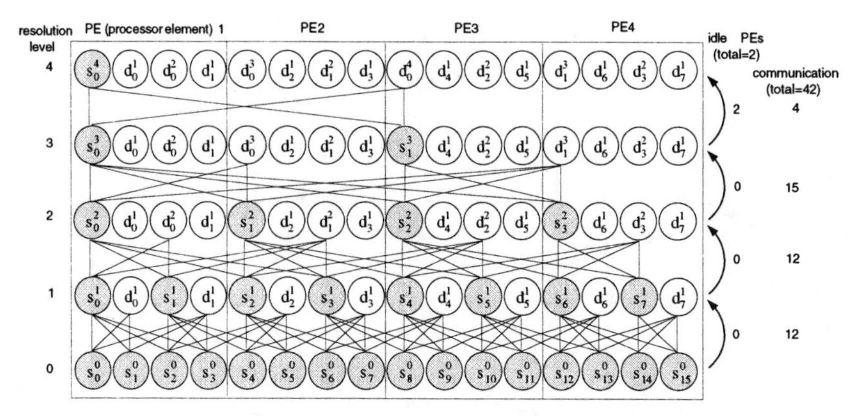

Figure 3.7. *Modified data arrangement in iDWT.*

Table 3.7. *Scalability vs. parallel efficiency (parallel efficiency in parentheses).*

PEs	SOR (Red-Black)	CG	original iDWTCG	modified iDWTCG
4	4.0 (100)	4.0 (100)	2.3 (58)	4.0 (100)
16	16.0 (100)	15.9 (99)	6.4 (40)	15.9 (99)
64	63.2 (97)	61.8 (97)	19.4 (30)	59.2 (93)
128	105.5 (82)	89.9 (70)	34.1 (27)	101.9 (80)

cessing since it leads to a great deal of processor idling. We modify the data arrangement to that shown in Figure 3.7, to reduce the number of idle processors, by more evenly balancing computational loads and reducing data communication overhead. Results from using our new data arrangement with 512×512 grid points (shown in Figure 3.8) show that the computing time of the iDWTCG method decreases at the same rate as the other methods. It is also worth noting that solving a 1024×1024 size problem with iDWTCG takes only the same order seconds as solving a 512×512 problem with SOR or CG.

Table 3.7 shows parallel scalability versus efficiency for the SOR, CG, and iDWTCG methods with and without a modified data arrangement, using 512×512 grid points. Our modified data arrangement enables users to take advantage of the improved computational performance from iDWTCG preconditioning for large-scale computations.

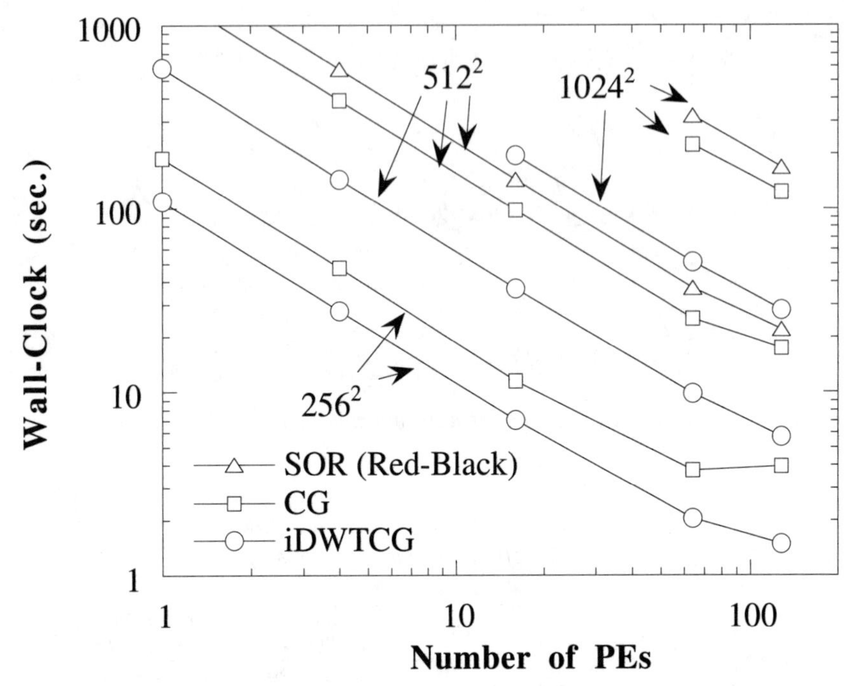

Figure 3.8. *Parallel performance of modified iDWTCG solver.*

3.4. Conclusion

In this chapter, we proposed a new preconditioning method called the incomplete discrete wavelet transform (iDWT) and applied it to solve Poisson equations. iDWT preconditioning improves the condition number of linear systems by shifting the eigenvalues of the coefficient matrix towards medium range values. The method is highly suitable for use prior to the application of iterative, projection-type matrix solvers, such as the CG method. The performance of the iDWTCG method becomes increasingly better than that of the ICCG method as the size of a problem becomes larger (for example, when successively finer meshes are used). Also, we demonstrated that our preconditioning scheme is equally effective for solving multidimensional problems.

Since our iDWTCG algorithm consists solely of local operations, it achieves high computing performance on vector and parallel computers. For example, on a Cray C94D vector computing system, the iDWTCG method can solve two-dimensional, discretized Poisson equations using 1024×1024

grid points about 14 times faster than the ICCG method. Our iDWTCG method is a highly efficient preconditioning method for CG when the code is tuned to run on parallel processing machines, because it contains much fewer data recurrence relations than incomplete Cholesky (IC) preconditioning. In fact, IC preconditioning contains so many data recurrence relations that its efficiency drops to a level which is worse than a system without preconditioning [18]. The iDWTCG method also performs well on a Cray T3D parallel computer; speedup by a factor of 100 is achieved by using 128 processors.

Appendix 1. Conjugate Gradient Methods

Conjugate gradient (CG) methods are used to determine a numerical approximation to the solution $\boldsymbol{x}$ of a system of n simultaneous equations

$$\boldsymbol{A}\boldsymbol{x} = \boldsymbol{b}; \qquad \boldsymbol{A} \in \boldsymbol{R}^{n\times n}, \quad \boldsymbol{x} \in \boldsymbol{R}^{n}, \quad \boldsymbol{b} \in \boldsymbol{R}^{n},$$

by computing successively better trial positions in n-dimensional space which minimize the error function. Let $\boldsymbol{x}_k$ be a trial solution to the system above. The associated residual vector is

$$\boldsymbol{r}_k = \boldsymbol{b} - \boldsymbol{A}\boldsymbol{x}_k.$$

If $\boldsymbol{A}$ is positive definite, then so is its inverse $\boldsymbol{A}^{-1}$, and

$$h = \boldsymbol{r}_k^T \boldsymbol{A}^{-1} \boldsymbol{r}_k > 0 \quad \text{for} \quad \boldsymbol{x}_k \neq \boldsymbol{x}.$$

(Note that $h = 0$ when $\boldsymbol{x}_k = \boldsymbol{x}$.) We substitute the expression for $\boldsymbol{r}_k$ to obtain

$$h = \boldsymbol{x}_k^T \boldsymbol{A} \boldsymbol{x}_k - 2\boldsymbol{b}^T \boldsymbol{x}_k + \boldsymbol{b}^T \boldsymbol{A}^{-1} \boldsymbol{b}.$$

Given an approximation $\boldsymbol{x}_k$, we will find a better approximation $\boldsymbol{x}_{k+1}$ by setting

$$\boldsymbol{x}_{k+1} = \boldsymbol{x}_k + \alpha \boldsymbol{p}_k$$

for a user-specified vector $\boldsymbol{p}_k$. (We will discuss methods for selecting $\boldsymbol{p}_k$ in the following paragraphs.) Substitution of this expression for $\boldsymbol{x}_{k+1}$ into the equation for h yields

$$h = \alpha^2 \boldsymbol{p}_k^T \boldsymbol{A} \boldsymbol{p}_k - 2\alpha \boldsymbol{p}_k^T \boldsymbol{r} + \boldsymbol{x}_k^T \boldsymbol{A} \boldsymbol{x}_k + \boldsymbol{b}^T \boldsymbol{A}^{-1} \boldsymbol{b}.$$

A local minimum of h (as a function of α) can be determined by setting the first differential to zero, i.e.,

$$\frac{\partial h}{\partial \alpha} = 2\boldsymbol{p}_k^T \left[\alpha \boldsymbol{A}_k - \boldsymbol{r}_k\right] = 0.$$

From the equation above, we determine α which minimizes h and use the local minimum as the new trial vector $\boldsymbol{x}_{k+1}$ in the next iteration step, i.e., we set

$$\boldsymbol{x}_{k+1} = \boldsymbol{x}_k + \alpha_k \boldsymbol{p}_k,$$

where

$$\alpha_k = \frac{\boldsymbol{p}_k^T \boldsymbol{r}_k}{\boldsymbol{p}_k^T \boldsymbol{A} \boldsymbol{p}_k}.$$

The method of steepest descent and the CG method are based on this same approach described above; however, they use different choices for the vector $\boldsymbol{p}_k$. In the method of steepest descent, $\boldsymbol{p}_k$ points in the direction which maximizes the gradient of the error function h at $\boldsymbol{x}_k$. In the CG method, $\boldsymbol{p}_k$ are chosen to be in the direction which minimizes the error function h as much as possible, while satisfying the condition

$$\boldsymbol{p}_i^T \boldsymbol{A} \boldsymbol{p}_j = 0 \quad \text{for} \quad i \neq j.$$

The algorithm for the CG method is given below. Further details on the method can be found in [10] and [12].

Initialization.

(a) initial guess $\boldsymbol{x}_0$

(b) $\boldsymbol{r}_0 = \boldsymbol{b} - \boldsymbol{A}\boldsymbol{x}_0$

(c) $\boldsymbol{p}_1 = \boldsymbol{r}_0$

Iterative procedure.

(d) $\alpha_k = \dfrac{(\boldsymbol{r}_k, \boldsymbol{r}_k)}{(\boldsymbol{p}_k, \boldsymbol{A}\boldsymbol{p}_k)}$

(e) $\boldsymbol{x}_{k+1} = \boldsymbol{x}_k + \alpha_k \boldsymbol{p}_k$

(f) $\boldsymbol{r}_{k+1} = \boldsymbol{r}_k - \alpha_k \boldsymbol{A}\boldsymbol{p}_k$

(g) check for convergence

(h) $\beta_k = \dfrac{(\boldsymbol{r}_{k+1}, \boldsymbol{r}_{k+1})}{(\boldsymbol{r}_k, \boldsymbol{r}_k)}$

(i) $\boldsymbol{p}_{k+1} = \boldsymbol{r}_{k+1} + \beta_k \boldsymbol{p}_k$

Appendix 2. Preconditioned Conjugate Gradient Methods

Preconditioned conjugate gradient (PCG) methods are used to determine a numerical approximation to the solution $\boldsymbol{x}$ of a system of n simultaneous equations

$$\boldsymbol{A}\boldsymbol{x} = \boldsymbol{b}; \qquad \boldsymbol{A} \in \boldsymbol{R}^{n\times n}, \quad \boldsymbol{x} \in \boldsymbol{R}^{n}, \quad \boldsymbol{b} \in \boldsymbol{R}^{n},$$

by transforming the system to one which is easier to solve through a linear transformation, i.e., by applying a *preconditioning matrix* K to both sides of the system. Two main types of preconditioning matrices are used in PCG algorithms: incomplete Cholesky (IC) preconditioners [8] and polynomial preconditioners [1], [13], [16], [20].

In the PCG method, both sides of the system are multiplied by a preconditioning matrix $\boldsymbol{K}$. The matrix $\boldsymbol{K}$ is chosen to be positive definite and is one which reduces the condition number of $\boldsymbol{A}$,

$$\kappa\|\boldsymbol{A}\| \gg \kappa\|\boldsymbol{K}\cdot\boldsymbol{A}\| \ .$$

Since $\boldsymbol{K}$ is positive definite, the Cholesky decomposition theorem tells us that there exists a matrix $\boldsymbol{V}$ such that

$$\boldsymbol{K} = \boldsymbol{V}^{T}\boldsymbol{V}$$

[22]. Loosely speaking, we can think of positive-definite matrices as having a matrix square root. The system above is transformed by the matrix $\boldsymbol{V}$ to

$$\hat{\boldsymbol{A}}\,\hat{\boldsymbol{x}} = \hat{\boldsymbol{b}},$$

where

$$\begin{aligned}\hat{\boldsymbol{A}} &= \boldsymbol{V}\boldsymbol{A}\boldsymbol{V}^{T},\\ \hat{\boldsymbol{x}} &= \boldsymbol{V}^{-T}\boldsymbol{x},\\ \hat{\boldsymbol{b}} &= \boldsymbol{V}\boldsymbol{b}.\end{aligned}$$

To solve this new linear system, just apply the CG algorithm. A summary of the PCG algorithm is given below. One of the most popular preconditioners is the IC. In the IC preconditioning, the original matrix $\boldsymbol{A}$ is incompletely decomposed into

$$\boldsymbol{A} = \boldsymbol{U}^{T}\boldsymbol{U} + \boldsymbol{R},$$

where $\boldsymbol{U}$ is upper triangular, that is, as sparse as $\boldsymbol{A}$, and $\boldsymbol{R}$ is the residue matrix. The preconditioning matrix is chosen to be

$$\boldsymbol{K} = \boldsymbol{U}^{-1}\boldsymbol{U}^{-T} \cong \boldsymbol{A}^{-1}.$$

Further details can be found in [15].

Initialization.

(a) initial guess for $\boldsymbol{u}_0$
(b) $\boldsymbol{r}_0 = \boldsymbol{f} - \boldsymbol{A}\boldsymbol{u}_0$
(c) $\boldsymbol{p}_0 = \vec{\boldsymbol{r}}_0 = \boldsymbol{K}\boldsymbol{r}_0$

Iterative procedure.

(d) $\alpha_k = \dfrac{(\vec{\boldsymbol{r}}_k, \boldsymbol{r}_k)}{(\boldsymbol{p}_k, \boldsymbol{A}\boldsymbol{p}_k)}$
(e) $\boldsymbol{u}_{k+1} = \boldsymbol{u}_k + \alpha_k \boldsymbol{p}_k$
(f) $\boldsymbol{r}_{k+1} = \boldsymbol{r}_k - \alpha_k \boldsymbol{A}\boldsymbol{p}_k$
(g) Check for convergence
(h) $\vec{\boldsymbol{r}}_{k+1} = \boldsymbol{K}\boldsymbol{r}_{k+1}$
(i) $\beta_k = \dfrac{(\vec{\boldsymbol{r}}_{k+1}, \boldsymbol{r}_{k+1})}{(\vec{\boldsymbol{r}}_k, \boldsymbol{r}_k)}$
(j) $\boldsymbol{p}_{k+1} = \vec{\boldsymbol{r}}_{k+1} + \beta_k \boldsymbol{p}_k$

Appendix 3. Successive Over-Relaxation Methods

Successive over-relaxation (SOR) methods are used to determine solutions to systems of linear equations of the form

$$\boldsymbol{A}\boldsymbol{x} = \boldsymbol{b}; \qquad A \in \boldsymbol{R}^{n\times n}, \quad \boldsymbol{x} \in \boldsymbol{R}^n, \quad \boldsymbol{b} \in \boldsymbol{R}^n,$$

in which the matrix $\boldsymbol{A}$ has a strong leading diagonal. Note that systems can be scaled to have diagonal elements which are unity so that we can assume the system can be expressed

$$\boldsymbol{A} = \boldsymbol{I} - \boldsymbol{L} - \boldsymbol{U},$$

where $\boldsymbol{L}$ is a strictly lower triangular matrix, $\boldsymbol{I}$ is the identity matrix, and $\boldsymbol{U}$ is a strictly upper triangular matrix. SOR methods are a generalization of the Gauss–Seidel (GS) iterative method, which computes successively better approximations of the solution $\boldsymbol{x}$ through the iterative scheme

$$(\boldsymbol{I} - \boldsymbol{L})\ \boldsymbol{x}_{k+1} = b + \boldsymbol{U}\boldsymbol{x}_k.$$

Here $\boldsymbol{x}_{k+1}$ and $\boldsymbol{x}_k$ denote the approximations for $\boldsymbol{x}$ at the $(k+1)$st and kth iterative steps. The equation is derived from adding $\boldsymbol{U}\boldsymbol{x}$ to both sides of the expression

$$\boldsymbol{A}\boldsymbol{x} = (\boldsymbol{I} - \boldsymbol{L} - \boldsymbol{U})\boldsymbol{x} = \boldsymbol{b}_0.$$

Usually, the exact solution $\mathbf{x}$ cannot be numerically determined; however, a good approximation can. An example of a termination criterion for the GS iteration is

$$\frac{\|\boldsymbol{x}_k - \boldsymbol{x}_{k+1}\|}{\|\boldsymbol{x}_k\|} \leq \text{tolerance}$$

for a user-specified vector norm and tolerance parameter. Note that if the tolerance is set to 1 and $\boldsymbol{x}_0 - 0$, the criterion is automatically satisfied. Typical values used for the tolerance parameter in the GS method are on the order of 10^{-3}. The residual $\boldsymbol{r}_k$ defined by

$$\boldsymbol{r}_k = \boldsymbol{A}\boldsymbol{x}_k - \boldsymbol{b},$$

for the kth iteration step, is sometimes used as an alternative criterion for termination of the iteration. The GS method suffers from a slow rate of convergence in many applications so that introduction of an acceleration parameter is recommended. The SOR method modifies the GS iteration by scaling the term $\boldsymbol{x}_{k+1} - \boldsymbol{x}_k$ by an over-relaxation (or acceleration) parameter ω. Two successive SOR iterative steps are related by

$$(\boldsymbol{I} - \omega\boldsymbol{L})\ \boldsymbol{x}_{k+1} = \omega b + [\ (1-\omega)\boldsymbol{I} + \omega\boldsymbol{U}\ ]\ \boldsymbol{x}_k,$$

where ω is usually set within the range $1 < \omega < 2$. Note that the choice $\omega = 1$ yields the GS method. Further details on SOR methods can be found in [12], [21], and [23].

Acknowledgements. The author would like to thank Mei Kobayashi for her support in reviewing and revising the original manuscript.

References

[1] L. Adams (1985), "m-step preconditioned conjugate gradient methods," *SIAM J. Sci. Statist. Comput.*, vol. 6, pp. 452–463.

[2] G. Beylkin (1993), "On wavelet-based algorithms for solving differential equations," pp. 449–466 in *Wavelets: Mathematics and Applications*, J. Benedetto, M. Frazier (eds.), CRC Press, Tokyo.

[3] W. Dahmen, A. Kurdila, P. Oswald (eds.) (1997), *Multiscale Wavelet Methods for Partial Differential Equations*, Academic Press, Tokyo.

[4] I. Daubechies (1988), "Orthonormal bases of compactly supported wavelets," *Comm. Pure Appl. Math.*, vol. 41, pp. 909–996.

[5] I. Daubechies (1992), *Ten Lectures on Wavelets*, SIAM, Philadelphia, PA.

[6] J. Dongarra, I. Duff, D. Sorensen, H. van der Vorst (1991), *Solving Linear Systems on Vector and Shared Memory Computers*, SIAM, Philadelphia, PA.

[7] I. Duff, G. Meurant (1989), "The effect of ordering on preconditioned conjugate gradients," *BIT*, vol. 29, pp. 635–637.

[8] G. Golub, C. Van Loan (1989), *Matrix Computations*, Johns Hopkins Univ. Press, Baltimore, MD.

[9] R. Glowinski et al. (1990), "Wavelet solution of linear and non-linear elliptic, parabolic and hyperbolic problems in one space dimension," pp. 55–120 in *Computing Methods in Applied Science and Engineering*, R. Glowinski (ed.), SIAM, Philadelphia, PA.

[10] M. Hestenes, E. Steifel (1962), "Methods of conjugate gradient for solving linear systems," *J. Res. Nat. Bureau Standards*, vol. 49, pp. 409–436.

[11] S. Jaffard, Ph. Laurençot (1992), "Orthonormal wavelets, analysis of operators and applications to numerical analysis," pp. 543–601 in *Wavelets: A Tutorial and Applications*, C. Chui (ed.), Academic Press, Tokyo.

[12] A. Jennings, J. McKeown (1992), *Matrix Computation*, John Wiley, New York.

[13] O. Johnson, C. Micchelli, G. Paul (1983), "Polynomial preconditioning for conjugate gradient calculations," *SIAM J. Numer. Anal.*, vol. 20, pp. 362–376.

[14] T. Koornwinder (1993), "Fast wavelet transforms and Calderón Zygmund operators," pp. 161–182 in *Wavelets: An Elementary Treatment of Theory and Applications*, T. Koornwinder (ed.), World Scientific, Singapore.

[15] J. Meijerink, H. van der Vorst (1977), "An iterative solution method for linear systems for which the coefficient matrix is a symmetric M-matrix," *Math. Comp.*, vol. 31, pp. 148–162.

[16] Y. Saad (1985), "Practical use of polynomial preconditionings for the conjugate gradient method," *SIAM J. Sci. Statist. Comput.*, vol. 6, pp. 865–881.

[17] N. Tanaka et al. (1996), "Incomplete discrete wavelet transform and the application to a Poisson equation solver," *J. Nucl. Sci. Technol.*, vol. 33, pp. 555–561.

[18] N. Tanaka, H. Terasaka (1993), "A numerical study of incompressible viscous flow with massively parallel processors," *Comp. Fluid Dynamics J.*, vol. 2, pp. 145–160.

[19] N. Tanaka, H. Terasaka (1994), "Numerical analysis with wavelets," *Proc. Japanese Soc. Mech. Engineers 7th Comp. Mech. Conf.*, 940-954, pp. 91–92 (in Japanese).

[20] N. Tanaka, H. Terasaka (1994), "Development of a preconditioning method for conjugate gradient algorithms using polynomial of a matrix blocked in small scale," *Trans. Inform. Process. Soc. Japan*, vol. 35, pp. 1519–1530 (in Japanese).

[21] R. Varga (1962), *Matrix Iterative Analysis*, Prentice–Hall, Englewood Cliffs, NJ.

[22] D. Watkins (1991), *Fundamentals of Matrix Computations*, John Wiley, New York.

[23] D. Young (1950), *Iterative Methods for Solving Partial Differential Equations of Elliptic Type*, Ph.D. Thesis, Harvard Univ., Cambridge, MA.

4. Wavelet Analysis for a Text-to-Speech (TTS) System

Mei Kobayashi,* Masaharu Sakamoto,* Takashi Saito,*
Yasuhide Hashimoto,* Masafumi Nishimura,*
and Kazuhiro Suzuki*

Abstract. This chapter presents results from a case study in which wavelet analysis was used in the development of a Japanese text-to-speech (TTS) system for personal computers (PCs). We developed two new wavelet-based technologies to improve the quality of speech output from TTS systems: accurate pitch mark determination by wavelet analysis and speech waveform generation using a modified time domain pitch-synchronous overlap-add (TD-PSOLA) method.

Key words. text-to-speech, speech synthesis, wavelets, wavelet transform, time domain pitch synchronous overlap-add

4.1. Introduction

This chapter presents results from a case study in which wavelet analysis was used in the development of a Japanese text-to-speech (TTS) system for personal computers (PCs). Use of TTS systems has become fairly common in Japan since advances in computer technology in the past decade have made it possible for even small PCs to run TTS software in real time. And recently, they are usually included in complimentary software tool kits for customers purchasing moderate to expensive hardware and multimedia systems.

---

*IBM Japan, Ltd., Tokyo Research Laboratory, 1623-14, Shimotsuruma, Yamato-shi, Kanagawa-ken 242-8502 Japan (mei, sakamoto, saito, hasimoto, nisimura, suzukik@trl.ibm.co.jp). Portions of this case study appeared in "Wavelets for a Modern Muse" in the May 1996 issue of *SIAM News* and IBM Research Report RT0110, IBM Tokyo Research Lab, 1995 [45].

The naturalness of the synthesized speech is one of the most important features which can contribute to the success or demise of any TTS product since comprehensibility is, for most purposes, a resolved issue. Traditional synthesis methods, which are based on manipulation of speech signal spectrum (e.g., linear predictive coding synthesis and formant synthesis) produce comprehensible but unnatural-sounding output. The lack of naturalness commonly associated with these methods results from the use of over-simplified speech models, small synthesis unit inventories, and poor handling of text parsing for prosody control. We developed four new technologies to overcome these difficulties and improve the quality of output from TTS systems: accurate pitch mark determination by wavelet analysis, speech waveform generation using a modified time domain pitch synchronous overlap-add method, speech synthesis unit selection using a context-dependent clustering method, and efficient prosody control using a three-phrase parser. These technologies are used in a Japanese TTS system for PCs, and for ProTALKER version 1.0, a product of IBM Japan, Ltd. This paper will focus on the role of wavelet analysis in the development of the first two technologies.

This chapter is organized as follows. The remainder of this section reviews wavelets and their properties which are relevant to our work. In the second section, we discuss preliminary studies with wavelets (by other scientists as well as ourselves) which indicated their usefulness as a tool for acoustical analysis. TTS conversion and synthesis unit dictionary preparation are described in the third section. Since both of these tasks must be performed in the development and implementation of virtually all TTS systems, the details within distinguish the different TTS systems and their qualities. Two of our new technologies, wavelet analysis for pitch marking and a context-dependent clustering method for synthesis unit construction, are used for TTS dictionary preparation. Our three-phrase parser and overlap-add techniques are used for speech synthesis. Wavelet-based techniques will be emphasized in our presentation; however, other technologies will be sketched for the sake of maintaining the flow in presenting an overview of the system; references for further study are given for interested readers. Our conclusions and remarks on further directions for study are given in the fourth section.

4.1.1. Wavelets: Definition and Properties

Wavelets are families of functions $\psi_{a,b}$:

$$\psi_{a,b} \;=\; |a|^{-1/2}\,\psi\left(\frac{t-b}{a}\right); \qquad a,b \in \mathcal{R}, \quad a \neq 0$$

generated from a single function ψ *by dilations and translations* [6], [8]. One of the applications of this theory is to construct a basis set $\{\psi_{a,b}\}$ for efficient and accurate approximation of signals. The parameters a and b are often restricted to a discrete sublattice, i.e., the dilation step $a_0 > 1$ and translation step $b_0 \neq 0$ are fixed; then the corresponding wavelet family is

$$\psi_{m,n}(t) \;=\; |a_0|^{-m/2}\,\psi(a_0^{-m}t - nb_0),$$

where $a = a_0^m$ and $b = nb_0a_0^m$ [7], [8]. If the translation parameter b_0 is small, then the basis elements lie closer together, and the associated transforms are of a finer resolution.

Transforms are defined and used in wavelet methods to analyze signals in an analogous manner as Fourier methods [5], [22], [35], [41]. The continuous wavelet transform (CWT) for $f \in L^2(\mathcal{R})$ is defined as

$$(Uf)(a,b) \;=\; \langle \psi_{a,b}, f\rangle \;=\; |a|^{-1/2}\int dt\cdot\psi\left(\frac{t-b}{a}\right)\cdot f(t) \qquad (4.1)$$

for $a, b \in \mathcal{R}$, $a \neq 0$, and the discrete wavelet transform (DWT) as

$$(Tf)_m \;=\; \langle \psi_{m,n}, f\rangle \;=\; |a_0|^{-m/2}\int dt\cdot\psi(a_0^{-m}t - nb_0)\cdot f(t)$$

for $a_0 > 1$, $b_0 \neq 0$. Some good tutorials on wavelet transform analysis and implementation techniques are [1], [8], [30], [42], [48], and [56].

Computations of transforms using wavelets with compact support yield the same windowing effect as convolution with a time window function used in short-time Fourier spectral analysis; however, wavelet computations have a constant frequency-bandwidth ratio, or *constant-Q ratio.* Constant-Q analysis is well suited for speech signal analysis, because it allows very fine frequency resolution at low frequencies for coarse time grids and very fine temporal resolution at high frequencies. Speech signals are often classified into four categories: voiced sounds, fricatives, plosives, and silence. The sounds have different durations and frequency ranges. Wavelet transform (WT) is better than Fourier transform (FT) analysis for studying specific class types [16], [55]. In speech signal processing, sampling is often very fine so that we can approximate the CWT quite well by taking dt in equation (4.1) to be Δt, the length of the sampling interval. Since the signals we consider are one-dimensional and computations are simple, fast transform algorithms described in the literature, such as [7], [19], [17], [50], and [60], do not necessarily have to be used.

In another type of wavelet analysis of signals, instead of computing the wavelet, convolutions of a signal $\{f(j)\}$; $j = 0, 1, \ldots, N$ and wavelet filter bank coefficients $\{h(i)\}$; $i = 0, 1, \ldots, n$ can be used to analyze data, i.e., the

$$\sum_{i=1}^{n} h(i)\ f(n-i)$$

are computed and studied. A simple example of a wavelet filter bank computation with program code is given in [38], a thorough introductory discussion is presented in [53], and more advanced treatments are shown in [14], [58], and [59].

To analyze acoustical and speech signals, engineers often produce three-dimensional scalogram-like and phase-shift displays of WT data [36]; see, e.g., Figure 4.2. These displays, which are the analogues of those produced using FTs, are three-dimensional representations of speech signal spectra, with time represented on the x-axis and frequency on the y-axis. Gray scales are used to represent the third dimension, either the amplitudes (for scalograms) or the phase (for phase-shift diagrams), of the WT data. Normally, transforms are calculated for frequencies ranging between six octaves, with 12 halfsteps per octave, so that dyadic WT data are insufficient for producing useful scalograms and phase-shift displays. In short, over-complete sets of data are needed to produce these displays. An introductory discussion of speech signal processing which includes a discussion of visual processing of speech signals can be found in [39] and [41].

Early speculations on potential applications of WTs, based on simple scalogram and phase-shift displays of contrived, toy examples, such as the delta or step functions, have not necessarily been borne out. There are many types of wavelets, and it appears that success in using the method depends on the choice of the family and the associated computational algorithms. We examined a number of different wavelet families to determine those which might be suitable for meaningful speech signal analysis. These wavelets are described below.

4.1.2. Wavelets: Examples

In this section, we describe and show how to compute six types of wavelets with compact support which were used in our experiments: three different splines, Gabor, chirp, and the Mexican hat (see Figures 4.1a–f). We selected spline wavelets for our studies because they are well documented,

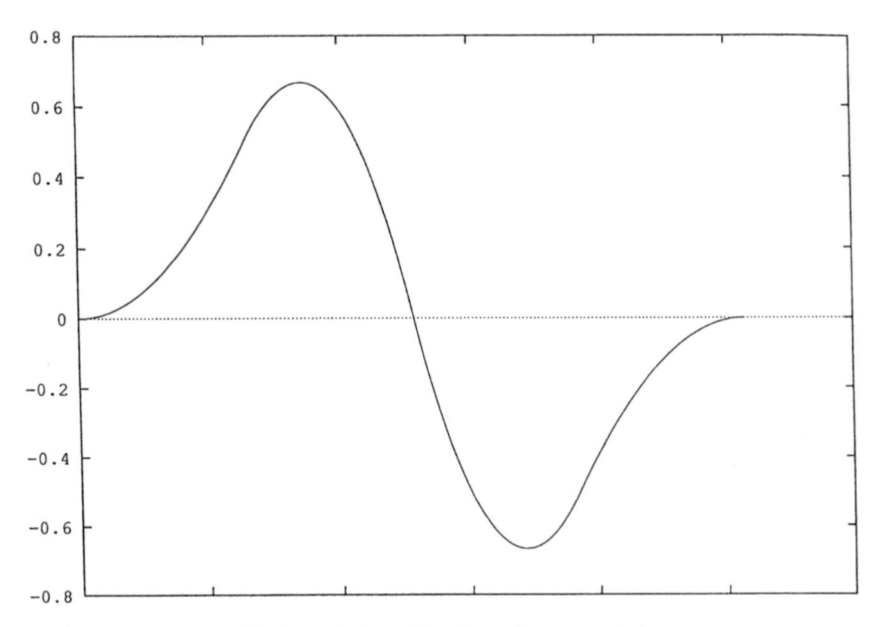

Figure 4.1a. *Simple spline wavelet.*

have a closed formula, and their applications are simple to implement. Gabor wavelet analysis showed great promise in frequency analysis of speech signals. We believed that chirp wavelet analysis could be used to detect changes in formant frequencies; however, our hopes were not borne out. Although the formula for the Mexican hat wavelet has an exponential term, the shape of the wavelet more closely resembles splines. We used the Mexican hat wavelet in our experiments to see if results would be closer to splines or the Gabor wavelet (which also has an exponential term) or would be in a class of its own. As mentioned earlier, wavelets defined solely by dyadic equations, such as orthonormal wavelets with compact support [6], were not considered because they do not yield enough data for producing meaningful scalograms and phase diagrams. Since speech data are very finely sampled, and the values of the wavelet functions can be accurately calculated on very small meshsizes, in our initial experiments we computed an approximation to the CWT given by equation (4.1). Fortunately, the number of nonzero values describing a particular wavelet are as large as several hundred, so that fine attention to handling of the end points was not needed since the values would be relatively small compared to central points and would not contribute significantly to the overall transform values.

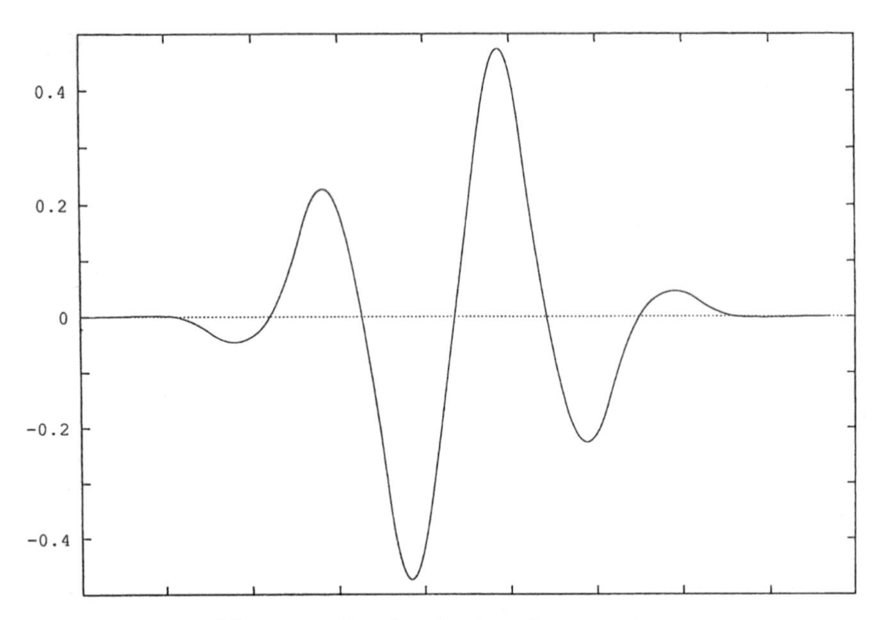

Figure 4.1b. *Quadratic spline wavelet.*

Splines have been used extensively to model functions [9], [37], [49] and, more recently, to model curves and surfaces in computer graphics [2], [3], [11], [51], [52]. (For references on wavelets and computer graphics, see section 1.1 of Chapter 1 of this book.) There are many types of splines, and they are generated by iteration or from explicit formulae. Although iterative methods tend to be computationally faster and more efficient, most algorithms are suited for calculations using 2^n evenly spaced sampling points, where $n \in \mathcal{Z}$; this only allows for octave-to-octave calculations. For these methods to be of practical value in speech processing, mathematical formulae or fast and accurate interpolation schemes are needed to determine wavelet function values for halfsteps between octaves. We used mathematical formulae to calculate splines for our experiments: a simple second-degree polynomial spline with continuous derivatives defined by

$$S(t) = \begin{cases} 2(t+1)^2, & -1 \le t < -\frac{1}{2}, \\ -6t^2 - 4t, & -\frac{1}{2} \le t < 0, \\ 6t^2 - 4t, & 0 \le t < \frac{1}{2}, \\ -2(t-1)^2, & \frac{1}{2} \le t < 1, \\ 0, & \text{otherwise,} \end{cases}$$

and slightly more sophisticated quadratic and cubic cardinal splines generated using the iterative formulae below [4]. First, the scaling functions are

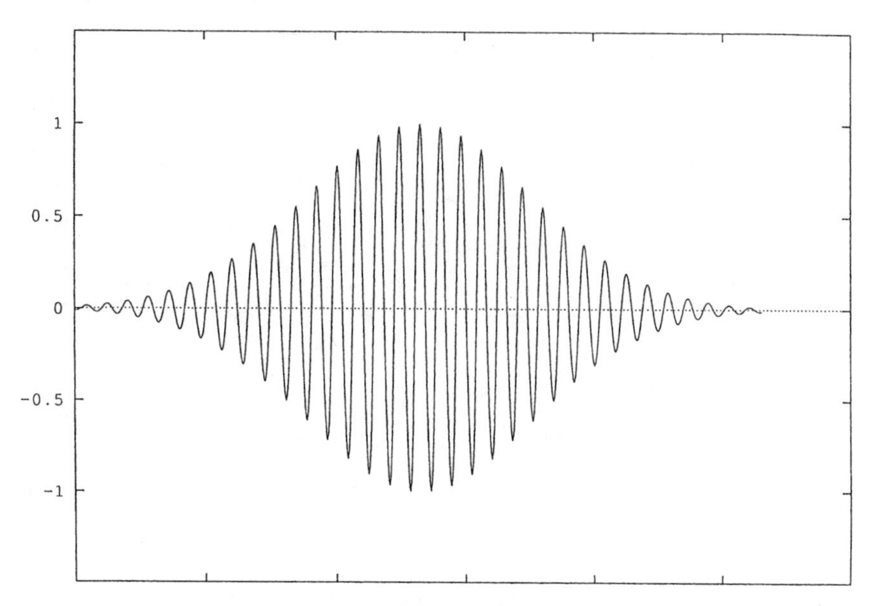

Figure 4.1c. *Cubic spline wavelet.*

computed from the formulae

$$N_1(t) = \chi_{[0,1[},$$

$$N_m(t) = \frac{t}{m-1} N_{m-1}(t) + \frac{m-t}{m-1} N_{m-1}(t-1),$$

where $\chi_{[0,1[}$ denotes the characteristic function on the semi-open unit interval. Then, the corresponding spline wavelets S_m are determined by

$$S_m(t) = \sum_n q_n N_m(2t-n);$$

$$q_n = \frac{(-1)^n}{2^{m-1}} \sum_{l=0}^{m} \binom{m}{l} N_{2m}(n+1-l), \quad n = 0, 1, \ldots, 3m-2.$$

Cardinal splines were selected because of their low order and odd and even properties. A filterbank approach to B-spline WT analysis is discussed in [57].

As the order of the cardinal B-splines is increased, the splines approach the Gabor wavelet

$$g_k(t) = \exp(-\alpha^2 t^2) \cdot \exp(j2\pi f_k t)$$

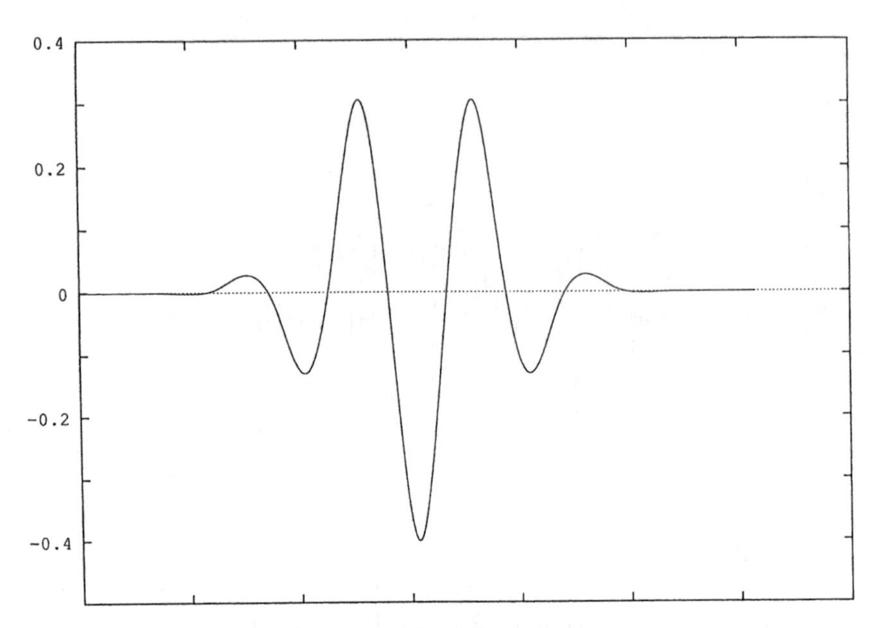

Figure 4.1d. *Gabor wavelet.*

as the limit function. As one would guess, Gabor wavelets are computed from this formula rather than iteratively from cardinal splines. Since the wavelets do not have compact support, during implementations, we set them to be zero in regions where the values are very small.

Gabor wavelets belong to a broader class of wavelets known as chirps, which are calculated from the formula

$$c_k(t) = \exp(-\alpha^2 t^2) \cdot \exp\left(j2\pi f_k t + \frac{1}{2} r t^2\right)$$

[12], [18], [28]. The frequency shift term $\frac{1}{2}rt^2$ is much smaller than $j2\pi f_k t$; the slight shift it causes in the period of the oscillations can be seen by comparing Figures 4.1d–e. Setting $r = 0$ yields the Gabor wavelet. Since chirps also do not have compact support, as with the Gabor wavelets, we set their values to be zero when the function values become very small. Further extensive discussion of chirps can be found in [28] and [29].

The last wavelet we considered is the Mexican hat:

$$M(t) = \frac{2}{\sqrt{3}} \pi^{-1/4} (1 - x^2) \, \exp\left(\frac{-t^2}{2}\right).$$

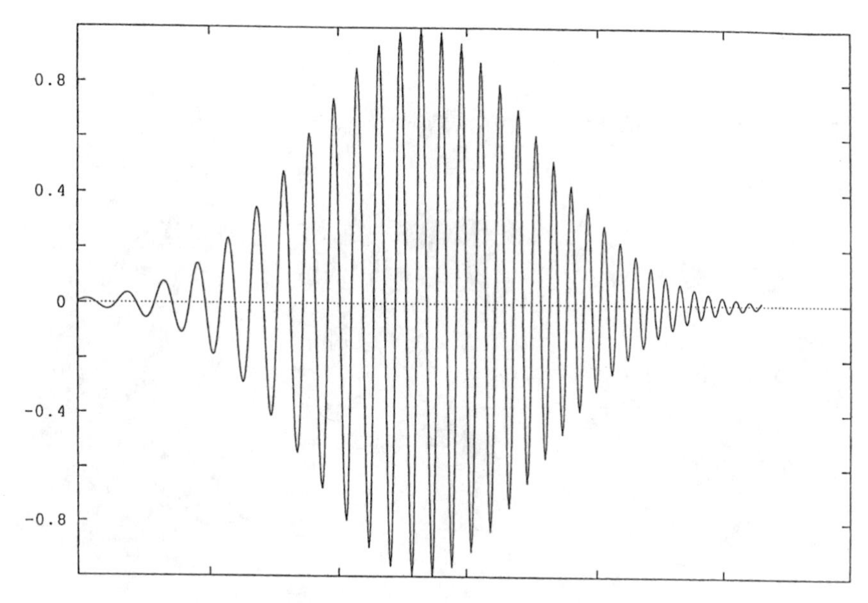

Figure 4.1e. *Chirp wavelet.*

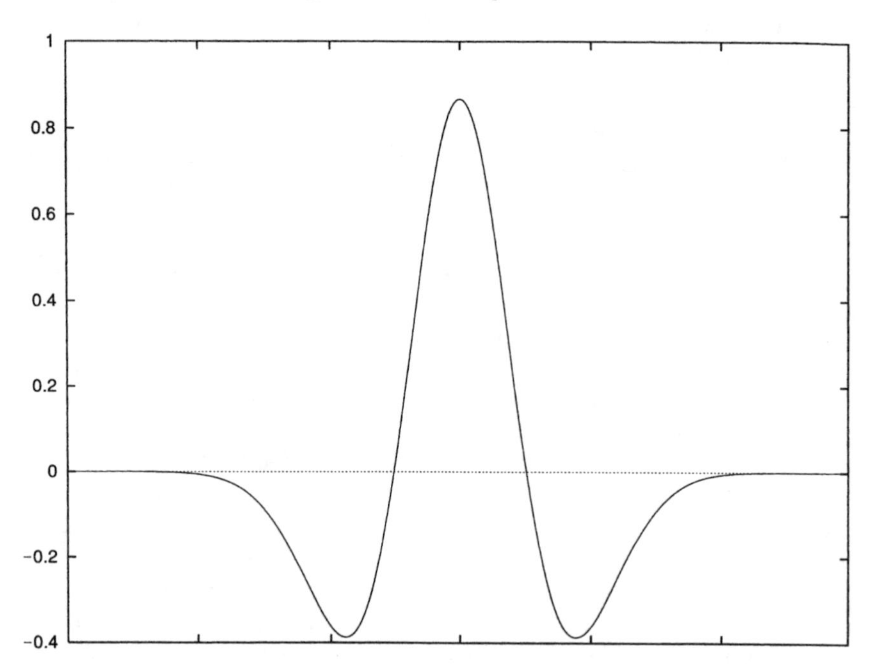

Figure 4.1f. *Mexican hat wavelet.*

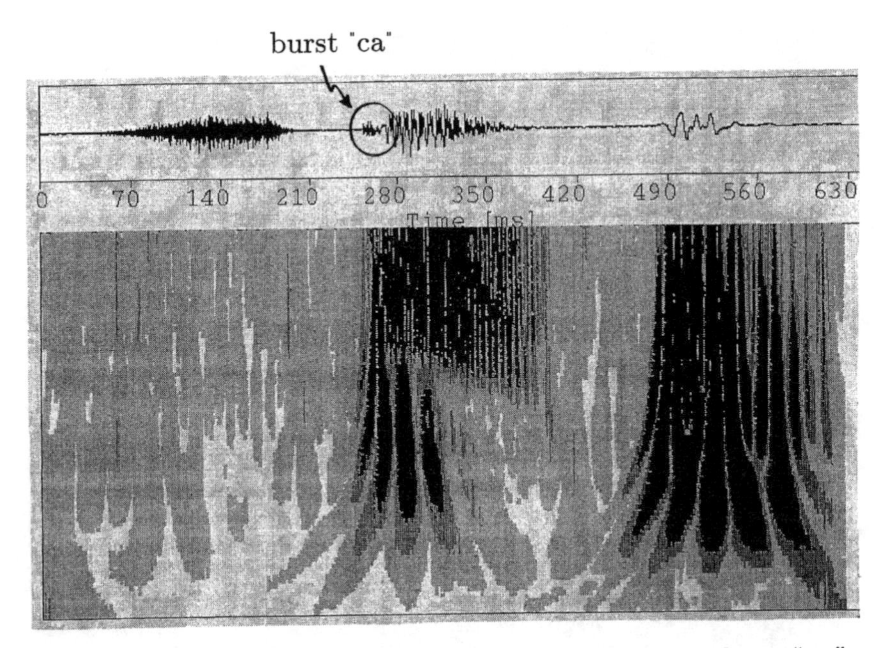

Figure 4.2. *Wavelet scalogram of "call up": splines detect burst "ca."*

The formula for the Mexican hat wavelet has an exponential term, which is the second derivative of the Gaussian function $\exp(-t^2/2)$. Again, these wavelets are set to have compact support during computations.

4.2. Wavelets and Speech Signal Processing

Our acoustical experiments were influenced by the work of many scientists, only a few of whom we mention below. Early acoustical signal processing work using wavelets is described in [25], [61], studies on the theory that wavelet methods are closely related to innate human sound processing mechanisms can be found in [18], [19], and the idea that wavelets might be useful for speaker and speech recognition is discussed in [21], [47]. More recent and specialized work with human speech and wavelets by Dorize and Gram-Hansen [10], Hiyane and Sawabe [16], Liénard and d'Alessandro [26], Sakamoto and Nishimura [46], and Tan et al. [55] shows promise in speech event and word boundary detection. In particular, the work by Tan et al. shows how to identify four categories of speech: voiced speech, plosives, fricatives, and silence; however, the data used were unusually clean and free of noise. However, the application targets of our experiments were different. We were looking for alternative techniques to use in speech synthesis/recognition systems; Tan et al. were seeking to develop improved hearing aid devices.

We conducted some experiments to confirm that similar results could be obtained using less clean data and a different set of wavelets [24] and to determine if the shape of the wavelet or the mathematical formula, e.g., the exponential term, would be important for speech feature identification. Six types of wavelets were considered: three splines, the Gabor, chirp, and the Mexican hat. The wavelet shape may be more important than the mathematical definition in determining scalogram features (i.e., scalograms for simple polynomial, quadratic, and cubic cardinal splines, and for the Mexican hat show similar overall patterns, and scalograms from the Gabor and chirp showed another type of similar pattern); splines and the Mexican hat appear to be the best for identifying the locations of bursts (see, e.g., Figure 4.2); and fricatives could be identified using one set of data from spline or Mexican hat analysis coupled with one set of data from Gabor or chirp analysis.

In a follow-up experiment, we examined whether WT-based speech recognition for the Japanese language is a realistic possibility. Very early results from phoneme recognition experiments using 5240 Japanese words (male voice) from the Advanced Telecommunications Research Institute (ATR) database MAU indicate a slightly worse rate for WT methods; however, the algorithm has been tuned for FT methods [53]. The extent of overlap in the error sets from the WT and FT methods must be determined to assess the usability of WTs in speech and/or phoneme recognition systems. Furthermore, modification of the current algorithm for FT methods must be made or a WT-specific algorithm developed.

In a separate follow-up experiment, we confirmed and enhanced results from the use of a new wavelet-based pitch-marking method reported by Kadambe and Boudreaux-Bartels [20]. Pitch, in human speech, is defined as the frequency of vocal cord vibration or the inverse of the time between glottal closure instants (GCIs)—the moments when the vocal cords close during speech. GCIs can serve as reference points for accurate pitch marking. Pitch information is vital for a number of applications, such as speech communication (analysis, transmission, and synthesis of speech as well as speech and speaker recognition); phonetics and linguistics (study of prosodic and phonetic features, such as tone, word stress, emotions); education (teaching the deaf intonation and students prosody of foreign languages); medicine (diagnosis of disease); and musicology [15]. A variety of pitch-marking techniques have been developed over the years [15], [41].

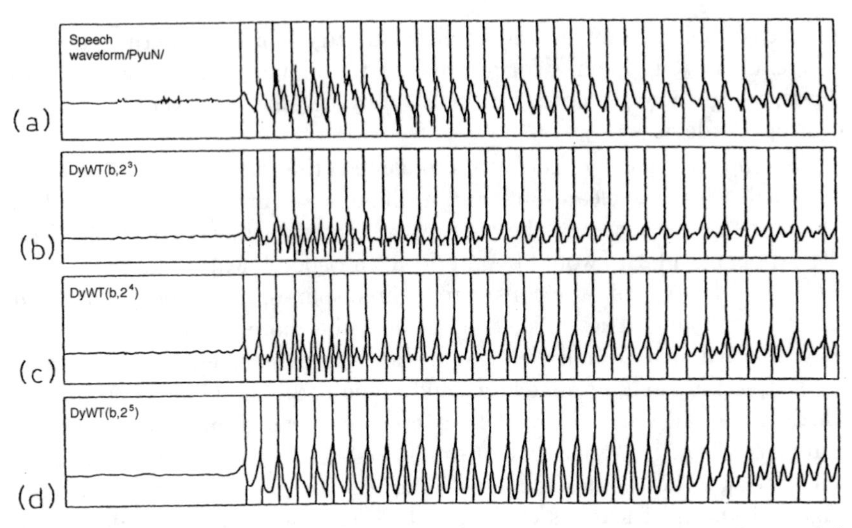

Figure 4.3. *Determination of pitch period using wavelet transforms.* (a) *original signal;* (b), (c), *and* (d) *WT of signal at successively coarser levels.*

In addition to the work by Kadambe and Boudreaux-Bartels, wavelet-based methods have been developed by Kawahara and Cheveigine [23] and Yip, Leung, and Wong [62].

We developed a program to detect GCIs by modifying the method of Kadambe and Boudreaux-Bartels which is based on the abrupt change produced by a GCI in a speech waveform. We examined wavelet transform methods because Kadambe and Boudreaux-Bartels showed, through several examples, that wavelet-based pitch detectors yield more superior results than classical pitch detectors which are based on autocorrelation and cepstrum methods. Classical pitch detectors compute the average pitch over a segment which covers several pitches. This pitch information is not suitable for our purposes since we need to know the pitch for each individual period. More recent techniques determine reference points for marking pitches by marking the maxima of major peaks in the speech waveform. However, these peaks are not perfectly smooth and have many local maxima so that accurate marking is difficult. Wavelet-based pitch marking overcomes this difficulty, because the wavelet-transformed data are smoother, and the location of the maximum is unambiguous. Additional reasons cited by Kadambe and Boudreaux-Bartels for the better performance by wavelet-based methods are that the pitch detector is suitable for a wide range of pitch periods; it can detect the beginning of a pitch period and the number of pitch periods present in a given segment of a speech signal; and it is computationally simple.

Kadambe and Boudreaux-Bartels' pitch detector computes the WT of an input speech signal; the cubic spline wavelet of Mallat and Zhong [27] yielded good results, and three dyadic scales were found to be sufficient for estimating the pitch period. We found that wavelets with the same overall shape as the reported cubic spline yielded equally good results.

Our pitch program detects GCIs by searching for a local peak in the wavelet transform of the speech waveform and by using two consecutive peaks to estimate the pitch period. An adaptive threshold uses dyadic wavelets to accurately and stably extract GCIs (see Figure 4.3). In contrast to Fourier transform methods, which require fine tuning for each speaker, our wavelet transform–based program appears to be more stable and robust for pitch marking both male and female voices. In some preliminary studies, our method has a 97% success rate for identification of GCIs [34]. Our findings need to be substantiated with further experiments. Results using more conventional methods are given in [31]. In the next section, we will see how our automatic pitch-marking tool contributed to the development of our TTS system. The pitches in all of the speech synthesis units installed in the TTS system were marked by our tool and were subsequently double-checked manually. The importance of accurate pitch marking for producing clear, high quality speech will become evident in our discussions.

4.3. TTS Conversion

The three main steps in the TTS process are illustrated in Figure 4.4. TTS conversion begins with a morphological analysis of a text file in Japanese characters. This step involves segmentation of the text into words, followed by grammatical context analysis of the words by means of a text analysis dictionary to determine the correct pronunciation of the characters, each of which usually stands for either one or two syllables in Japanese. This step requires a significant amount of work because written Japanese, unlike modern European languages (including English) does not have spaces between words, and because the pronunciation of a Japanese character usually varies with meaning and context.

In the second step, a three-phase parser is used to control prosody—the placement of accentuations, pauses, and other features of synthesized speech in the text. The parser examines each series of three consecutive "phrases" (syllables) A, B, and C. For a given syllable A, the system must select one of four relationships—accentuation, neutral connection, reset-

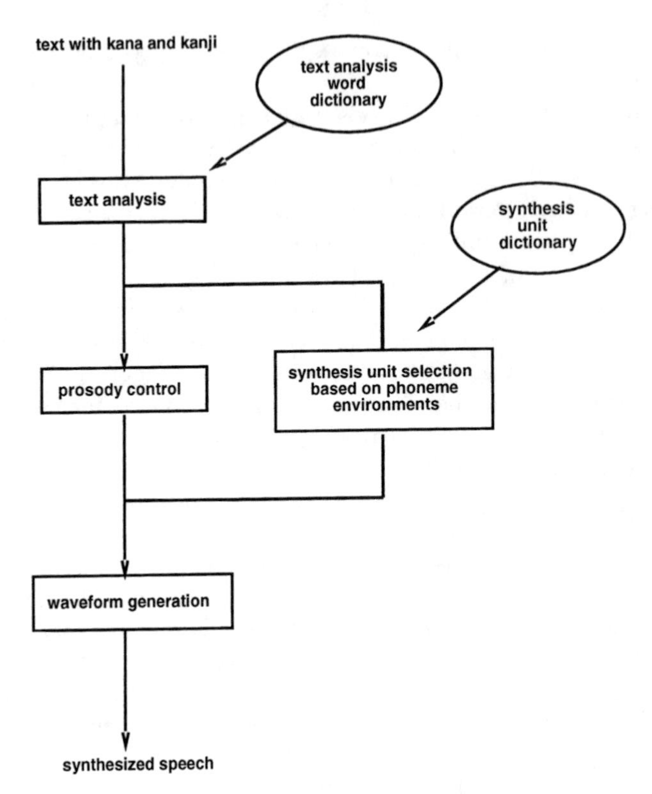

Figure 4.4. *Text-to-speech process.*

ting of pitch frequency, or pause—between the syllable and the subsequent syllable B. All combinations of pairs ($A - B$, $B - C$, and $A - C$) are also examined to determine whether or not a modifier-modifiee relationship exists.

We chose three-phrase parsing because, although simple, it gives good prosody commands for the Japanese language (but not necessarily for other languages). Parsers that examine a larger number of consecutive syllables must consider a greater number of phrase relationships, many of which lead to errors or ambiguities that are tedious to resolve. Details of the parsing method and studies of parsers that examine more than three phrases are discussed in [54].

The quality of synthesized speech produced by a TTS system depends strongly on the dictionary of synthesis units. Synthesis units can be thought of as speech sound templates, concatenations of which are used to produce speech. The need for a set of synthesis units large enough to produce high quality speech must be balanced with limits on the size of the dictionary.

Our method for selecting synthesis units to be installed in the dictionary is described in Appendix 1. A diagram summarizing the steps in the dictionary preparation process is given in Figure 4.5.

In the third and final step, the TTS system generates a speech waveform using synthesis units from the dictionary and the prosody control information determined in the second step. Generation of the waveform involves more than mere concatenation of appropriate synthesis units. The duration, pitch, and intensity—features that contribute to the naturalness of a synthesized voice—are all controlled by modifying the signal, i.e., the synthesis units, through a process known as time domain pitch-synchronous overlap-add (TD-PSOLA). The equation which serves as the basis for overlap-add (OLA) methods is derived in Appendix 2. The equation shows how a signal can be decomposed into short-term signals then recovered through an OLA process [33].

In pitch-synchronous overlap-add methods, first, the signal $s(n)$ is decomposed into short-term overlapping signals $s_m(n) = w_m(t_m - n) \cdot s(n)$, where $w_m(n)$ is an analysis window and t_m are pitch marks. These short-term signals are, in turn, modified to produce a new set of short-term signals $\tilde{s}_q(n)$, which are synchronized to have a new set of pitch marks $\tilde{t}_q$. Corresponding pitch marks t_m and $\tilde{t}_q$ are related by a time warping function. Speech is synthesized by overlap-adding the $\tilde{s}_q(n)$ using a synthesis window $\tilde{w}_q$ according to

$$\tilde{s}(n) = \frac{\sum_q \alpha_q \, \tilde{s}_q(n) \, \tilde{w}_q \left(\tilde{t}_q - n\right)}{\sum_q \tilde{w}_q^2 \left(\tilde{t}_q - n\right)}.$$

Our TTS system uses a modified version of an OLA method by Hamon, Moulines, and Charpentier [13] to produce smoother speech than other synthesis methods to date (see Figure 4.6). In Hamon's time domain pitch-synchronous overlap-add (TD-PSOLA) method, automatic pitch-synchronous editing of a speech signal is performed, that is, individual pitch periods are extracted then overlapped and added at a different rate to produce a synthesized speech signal. In this process, the original speech signal $s(n)$ is decomposed into a sequence of short-term, overlapping signals $s_m(n)$ using a *Hanning window*

$$w_H(n) = \begin{cases} \alpha + (1-\alpha)\cos(2\pi n/N); & |n| \leq (N-1)/2, \ \alpha = 0.5, \\ 0.0; & \text{elsewhere,} \end{cases}$$

centered around the time origin $n = 0$. And the short-term signals are given by $s_m(n) = w_H(T_m - n) \cdot s(n)$, where T_m denotes the pitch marks. Note that α must be in the range $0 \leq \alpha \leq 1.0$. When $\alpha = 0.5$, the window is a *Hanning window*, not to be confused with another common choice $\alpha = 0.54$,

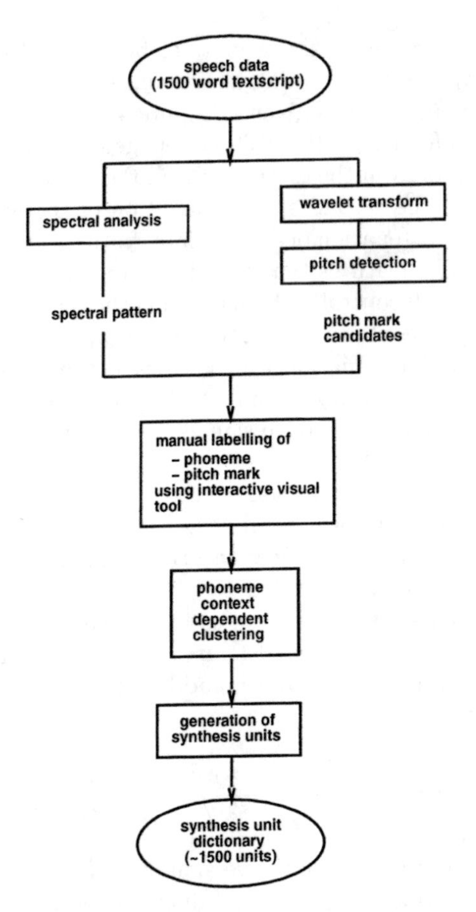

Figure 4.5. *Preparation of a synthesis unit dictionary.*

which yields a *Hamming window* [40]. The pitch marks T_m and the center of the Hanning window are set to coincide.

The primary difference in our method is a change in the relative placement of the synthesis window and the pitch marks. Pitch synchronization in the synthesis stage is completely independent of the placement of synthesis windows; the pitches in the speech synthesis units in our TTS system were marked using the GCI detection software tool described in the previous section, and windows are placed so as to minimize distortion in the synthesized speech. In our method (see Figure 4.6), glottal closure instants mark the beginning of pitch segments, and the maximum point of each pitch segment is used to align the center of a segmentation window. Since our wavelet-based method (described in section 4.2) stably and accurately determines pitch marks, rumbling sounds are minimized during synthesis.

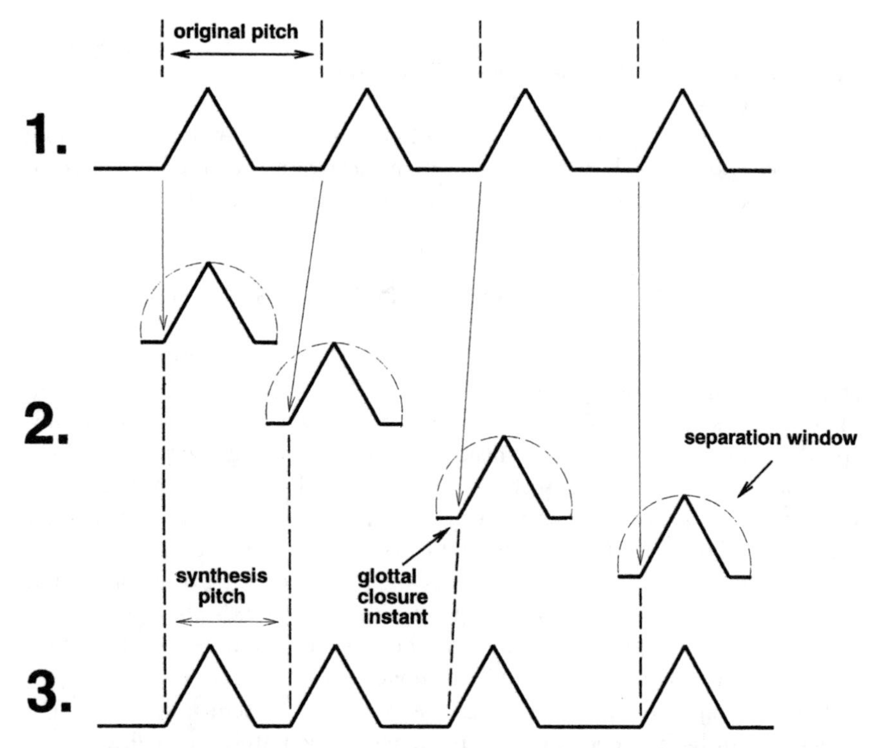

Figure 4.6. *Speech synthesis using a new overlap-add technique:* (1) *application of pitch mark;* (2) *waveform separation and overlapping by synthesis pitch;* (3) *synthesized waveform.*

4.4. Conclusion

In this chapter, we described two new wavelet-based technologies which were used to synthesize natural sounding speech in ProTALKER, a Japanese text-to-speech (TTS) system: accurate pitch mark determination by wavelet analysis and speech waveform generation using a modified time domain pitch-synchronous overlap-add method. The technologies improved the quality of synthesized speech.

Accurate pitch information is vital for a number of applications, including speech communication, phonetics and linguistics, education, medical diagnosis, and musicology. In addition to speech applications, there are many other types of signals (e.g., acoustical and nonacoustical, from physics and mechanics labs), which exhibit periodic or near periodic behavior. Our wavelet-based pitch-marking method may be a useful tool for investigating these research areas. Wavelet transform–based scalograms, which were used to detect transients in speech, are another set of useful tools which can and have been used to analyze one-dimensional signals in many different scientific areas.

Some emerging speech areas to explore using wavelet transforms and techniques are speech command system development or enhancement, speaker recognition, and medical diagnosis using speech data.

Appendix 1. Selection of Synthesis Units

The quality of synthesized speech produced by a TTS system depends strongly on the dictionary of synthesis units. Synthesis units can be thought of as speech sound templates, concatenations of which are used to produce speech. The need for a set of synthesis units large enough to produce high quality speech must be balanced with limits on the size of the dictionary. Phonemes, the sounds represented by consonants and vowels, are generally considered the smallest synthesis units of speech and also usually function as the smallest synthesis units in TTS systems. There are 38 basic phonemes in the Japanese language, but the quality of speech synthesized using only 38 units—one per phoneme—is very poor because it does not take into account the influence of surrounding phonemes.

The basic speech model we chose for our TTS system assumes that a phoneme can be influenced by the phoneme immediately preceding it and the one immediately following it; each phoneme can therefore have as many as 38^2 different environments. In actuality, not all 38^3 theoretically possible phonemes exist as sounds in the Japanese language, and the number of phonemes is further reduced when redundancies—for example, two and three consecutive pauses are the same as one—are taken into account. Still, there is not enough memory in many small PCs to install even the adjusted number of phonemes possible in this model, and owners of PCs that have large memories would not want a significant portion of their disk space to be tied up by one specialized software package.

To determine the set of units to install in the speech synthesis unit dictionary, we used a context-dependent clustering (CDC) algorithm developed by Hashimoto, Saito, and Sakamoto [16], [44] which enhances methods developed by Sagayama [43] and Nakajima [32]. In the CDC method, similar-sounding phonemes are grouped into clusters. The method determines how much distortion is introduced into the synthesized speech when one phoneme is used to represent all the phonemes in one cluster. If the distortion is small, then only one representative phoneme need be installed in the synthesis unit dictionary, yielding significant savings in memory. The CDC method can also be used as a guide for determining which set of phonemes would be the best (or near best) to install for a given amount of memory.

In the first step of the CDC algorithm, a metric is defined to allow the quantitative measurement of the distance (or difference in sound) between two phonemes. Then, for each cluster C, distances between all pairs of phonemes are calculated and used to determine the centermost—or most representative—phoneme, $P_{centroid}$:

$$P_{centroid} = P_i \in C \quad \text{which minimizes} \quad \sum_{j \in C} |P_i - P_j|; \qquad P_i, P_j \in C.$$

If the diameter of the cluster,

$$\text{diameter}(C) = \max_{P_i, P_j \in C} |P_i - P_j|,$$

is large, then at least two of the phonemes in C are not close, so the cluster should be split into two clusters with smaller diameters. Splitting the new clusters may, in turn, be necessary if the diameters are still too large. The CDC algorithm determines which clusters should be split and which phonemes should be assigned to which new cluster. Since quantitative analysis of the distortion depends on the accuracy of the speech model, which is not perfect, laboratory listening experiments are used to make final decisions during product development. All in all, however, the CDC algorithm is a useful tool for reducing the number of listening experiments that must be performed to select a good set of speech synthesis units.

We caution that, although our CDC algorithm is a useful tool for Japanese speech synthesis unit selection since the language is modeled fairly well by the basic, three-phoneme sequence model, extensions and variations of the algorithm must be developed for handling English and other languages which must be modeled using longer phoneme sequences.

Appendix 2: Overlap-Add Methods

In this appendix we derive the equation which serves as the basis for overlap-add (OLA) methods. The equation shows how a signal can be decomposed into short-term signals then recovered through an overlap-add process [33].

The *discrete Fourier transform (DFT)* for a set of data $\{x(n)\}$; $n \in \mathcal{Z}$ is defined as

$$X(\omega) = \sum_{n=-\infty}^{\infty} x(n) \, e^{-j\omega n},$$

where ω, the frequency, is a continuous variable. The discrete short-time Fourier transform (STFT) takes short-time windowed sections $w(n-m) \cdot x(n)$ of the original signal, i.e.,

$$X(\omega) = \sum_{m=-\infty}^{\infty} x(m)\, w(n-m)\, e^{-j\omega m},$$

where $w(n-m)$ is the *analysis filter* or *analysis window.*

The OLA method is based on the identity

$$x(n) = \frac{1}{2\pi W(0)} \int_{-\pi}^{\pi} \sum_{r=-\infty}^{\infty} X(r,\omega)\, e^{j\omega n}\, d\omega,$$

where

$$W(0) = \sum_{n=-\infty}^{\infty} w(n),$$

and can be derived from the equations above.

A signal can be synthesized using the discrete version of the integral above, i.e.,

$$x(n)_{synthesized} = \frac{1}{W(0)} \sum_{p=-\infty}^{\infty} \frac{1}{N} \sum_{k=0}^{N-1} X(p,k)\, e^{j2\pi kn/N}.$$

Since the inner sum is the inverse DFT for each p,

$$\begin{aligned} x(n)_{synthesized} &= \frac{1}{W(0)} \sum_{p=-\infty}^{\infty} x(n)\, w(p-n) \\ &= x(n) \cdot \frac{1}{W(0)} \sum_{p=-\infty}^{\infty} w(p-n). \end{aligned} \tag{4.2}$$

The synthesized and original signals are identical if

$$\sum_{p=-\infty}^{\infty} w(p-n) = W(0), \tag{4.3}$$

or, if

$$\sum_{p=-\infty}^{\infty} w(pL-n) = \frac{W(0)}{L},$$

where L denotes the decimation factor with respect to time. Substitution of the expression in (4.3) into equation (4.2) yields the basic synthesis equation for OLA methods,

$$x(n)_{synthesized} = \frac{L}{W(0)} \sum_{p=-\infty}^{\infty} \frac{1}{N} \sum_{k=0}^{N-1} X(pL,k)\, e^{j2\pi kn/N}.$$

Acknowledgements. The authors would like to thank Shubha Kadambe, Ryutarou Ohbuchi, and Yoshiaki Oshima for helpful discussions and Rena Bloom, Gail Corbett, and the staff at *SIAM News* for their kind encouragement and editorial help.

References

[1] A. Aldroubi (1996), "The wavelet transform: A surfing guide," pp. 3–36 in *Wavelets in Medicine and Biology*, A. Aldroubi, M. Unser (eds.), CRC Press, New York.

[2] B. Barsky (1988), *Computer Graphics and Geometric Modeling Using Beta-splines*, Springer-Verlag, Tokyo.

[3] B. Bartles, J. Beatty, B. Barsky (1987), *An Introduction to Splines for Use in Computer Graphics and Geometric Modeling*, Morgan-Kaufmann, San Francisco, CA.

[4] C. Chui (1992), *An Introduction to Wavelets*, Academic Press, Tokyo.

[5] R. Crochiere, L. Rabiner (1983), *Multirate Digital Signal Processing*, Prentice–Hall, Englewood Cliffs, NJ.

[6] I. Daubechies (1988), "Orthonormal bases of compactly supported wavelets," *Comm. Pure Appl. Math.*, vol. XLI, pp. 909–996.

[7] I. Daubechies (1990), "The wavelet transform, time-frequency localization and signal analysis," *IEEE Trans. Inform. Theory*, vol. 36, pp. 961–1005.

[8] I. Daubechies (1992), *Ten Lectures on Wavelets*, SIAM, Philadelphia, PA.

[9] C. deBoor (1978), *A Practical Guide to Splines*, Springer-Verlag, Tokyo.

[10] C. Dorize, K. Gram-Hansen (1992), "Related positive time-frequency energy distributions," and "On the choice of parameters for time-frequency analysis," pp. 77–92 in *Wavelets and Applications*, Y. Meyer (ed.), Springer-Verlag, Tokyo.

[11] G. Farin (1990), *Curves and Surfaces for Computer Aided Geometric Design*, Academic Press, Tokyo.

[12] A. Grossman, R. Kronland-Martinet, and J. Morlet (1989), "Reading and understanding continuous wavelet transforms," pp. 2–20 in *Wavelets: Time-Frequency Methods and Phase Space*, J. Combes, A. Grossmann, Ph. Tchamitchian (eds.), Springer-Verlag, Tokyo.

[13] C. Hamon, E. Moulines, and F. Charpentier (1989), "A diphone synthesis system based on time-domain prosodic modifications of speech," pp. 238–241 in *Proc. ICASSP*, IEEE Computer Society Press, Piscataway, NJ.

[14] C. Herley (1993), *Wavelets and Filter Banks*, Ph.D. Thesis, Columbia Univ., New York, NY.

[15] W. Hess (1983), *Pitch Determination of Speech Signals*, Springer-Verlag, Tokyo.

[16] K. Hiyane and N. Sawabe (1995), "Sound recognition using wavelet transform and genetic algorithm," in *Proc. Symp. Real World Computing*, Japanese Ministry of Internat. Trade and Industry, Tokyo, Japan; also available as: K. Hiyane, N. Sawabe (1994), *Proc. 5th Mtg. Shinkei Kairo Gakkai*, 34/35 (in Japanese).

[17] M. Holschneider, R. Kronland-Martinet, J. Morlet, and Ph. Tchamitchian (1988), "The Algorithme à Trous," CNRS Report CPT-88/P.2115, Centre National de la Recherche Scientifique, Paris, France. May.

[18] T. Irino (1992), "Speech signal processing using wavelet transform," *IEICE Technical Report*, Sp92-81, DSP92-66, Institute of Electronics, Information and Communication Engineers, Tokyo, Japan, Oct., pp. 59–66 (in Japanese).

[19] T. Irino, H. Kawahara (1993), "Signal reconstruction from modified auditory wavelet transform," *IEEE Trans. Signal Process.*, vol. 41, pp. 3549–3553.

[20] S. Kadambe, G. Boudreaux-Bartels (1991), "A comparison of wavelet functions for pitch detection of speech signals," pp. 449–452 in *Proc. ICASSP*, IEEE Computer Society Press, Piscataway, NJ.

[21] S. Kadambe, P. Srinivasan (1994), "Applications of adaptive wavelets for speech," SPIE *Opt. Engrg.*, vol. 33, pp. 2204–2211.

[22] J.-P. Kahane, P. Lemarié-Rieusset (1995), *Fourier Series and Wavelets*, Gordon and Breach, Amsterdam.

[23] H. Kawahara, A. Cheveigine (1996), "Error free F0 extraction method and its evaluation," *IEICE Technical Report*, SP96-96, 9/18, Institute of Electronics, Information and Communication Engineers, Tokyo, Japan (in Japanese).

[24] M. Kobayashi, M. Sakamoto (1993), "Wavelets and speech signal processing," Chapter 3, *Proc. SIAM Wavelet Seminar II*, JSIAM, Tokyo, Japan.

[25] R. Kronland-Martinet, J. Morlet, A. Grossman (1987), "Analysis of sound patterns through wavelet transforms," *Internat. J. Pattern Recognition*, vol. 1, pp. 273–302.

[26] J. Liénard, C. d'Alessandro (1989), "Wavelets and granular analysis of speech," pp. 158–163 in *Wavelets: Time-Frequency Methods and Phase Space*, J. Combes, A. Grossmann, Ph. Tchamitchian (eds.), Springer-Verlag, Tokyo.

[27] S. Mallat, S. Zhong (1992), "Characterization of signals from multiscale edges," *IEEE Trans. Pattern Anal. Mach. Intell.*, vol. 14, pp. 710–732.

[28] S. Mann, S. Haykin (1992), "Adaptive chirplet transform: An adaptive generalization of the wavelet transform," *SPIE Opt. Engrg.*, vol. 31, pp. 1243-1256.

[29] S. Mann, S. Haykin (1995), "The chirplet transform," *IEEE Trans. Signal Process.*, vol. 43, pp. 2745–2761.

[30] Y. Meyer (1990), *Ondelettes et Operateurs*, Hermann, Paris.

[31] E. Moulines et al. (1990), "A real-time French text-to-speech system generating high-quality synthetic speech," pp. 309–312 in *Proc. ICASSP*, IEEE Computer Society Press, Piscataway, NJ.

[32] S. Nakajima (1992), "English speech synthesis based on multi-level context oriented clustering method," *IEICE Speech Wkshp. Record*, SP92-9, Institute of Electronics, Information and Communication Engineers, Tokyo, Japan, May, pp. 17–24.

[33] S. Nawab, T. Quatieri (1988), "Short-time Fourier transform," pp. 289–337 in *Advanced Topics in Signal Processing*, J. Lim, A. Oppenheim (eds.), Prentice–Hall, Englewood Cliffs, NJ.

[34] M. Nishimura et al. (1994), "A study on speech compression and editing using continuous speech recognition system," *Proc. IPSJ Fall Meeting*, Information Processing Society of Japan, Tokyo, pp. 6G–10 (in Japanese).

[35] A. Oppenheim, R. Schafer (1989), *Discrete-Time Signal Processing*, Prentice–Hall, Englewood Cliffs, NJ.

[36] J. Pitton, K. Wang, B.-H. Juang (1996), "Time-frequency analysis and auditory modeling for automatic recognition of speech," *Proc. IEEE*, vol. 84, pp. 1199–1215.

[37] M. Powell (1981), *Approximation Theory and Methods*, Cambridge Univ. Press, Cambridge.

[38] W. Press, S. Teukolsky, W. Vetterling, and B. Flannery (1982), *Numerical Recipes in C*, 2nd ed., Cambridge Univ. Press, Cambridge.

[39] S. Qian, D. Chen (1996), *Joint Time-Frequency Analysis: Methods and Applications*, Prentice–Hall, Englewood Cliffs, NJ.

[40] L. Rabiner, B. Gold (1975), *Theory and Application of Digital Signal Processing*, Prentice–Hall, Englewood Cliffs, NJ.

[41] L. Rabiner, R. Schafer (1978), *Digital Processing of Speech Signals*, Prentice–Hall, Englewood Cliffs, NJ.

[42] O. Rioul, P. Duhamel (1992), "Fast algorithms for discrete and continuous wavelet transforms," *IEEE Trans. Inform. Theory*, vol. IT-38, pp. 569–586.

[43] S. Sagayama (1987), "Phoneme environment clustering, principle and algorithm," *IEICE Speech Wkshp. Record*, SP-87-86, Institute of Electronics, Information and Communication Engineers, Tokyo, Japan, Dec. 17.

[44] T. Saito, Y. Hashimoto, M. Sakamoto (1996), "High-quality speech synthesis using context-dependent syllabic units," pp. 381–384 in *Proc. ICASSP*, IEEE Computer Society Press, Piscataway, NJ.

[45] T. Saito et al. (1995), "ProTALKER: A Japanese text-to-speech system for personal computers," *IBM Research Report* RT0110, IBM Tokyo Research Lab, Japan, June.

[46] M. Sakamoto, M. Nishimura (1994), Speech recognition using wavelet transform, in *Proc. Semi-Annual Mtg., Acoust. Soc. Japan*, Spring, Vol. I, 1-Q-4 (poster session), pp. 117–118.

[47] T. Sakamoto and H. Tominaga (1993), "A study on speech recognition using wavelet transform," *Proc. of IEICE Semi-Annual Mtg.*, Spring, Institute of Electronics, Information and Communication Engineers, Tokyo, Japan, p. A-221 (in Japanese).

[48] M. Sato (1991), "Mathematical foundation of wavelets I and II," *J. Acoust. Soc. Japan*, vol. 47, pp. 405–423 (in Japanese).

[49] L. Schumaker (1981), *Spline Functions: Basic Theory*, John Wiley, New York, NY.

[50] M. Shensa (1992), "The discrete wavelet transform: Wedding the à trous and Mallat algorithms," *IEEE Trans. Signal Process.*, vol. 40, pp. 2464–2482.

[51] E. Shikin, A. Plis (1995), *Handbook on Splines for the User*, CRC Press, Tokyo.

[52] E. Stollnitz, T. DeRose, D. Salesin (1996), *Wavelets for Computer Graphics*, Morgan-Kaufmann, San Francisco, CA.

[53] G. Strang, T. Nguyen (1996), *Wavelets and Filter Banks*, Wellesley–Cambridge Press, Wellesley, MA.

[54] K. Suzuki, T. Saito (1995), "N-phrase parsing method for Japanese text-to-speech conversion and assignment of prosodic features based on N-phrase structures," *Trans. IEICE*, vol. J78 D-II, pp. 177–187.

[55] B. Tan et al. (1994), "Applying wavelet analysis to speech segmentation and classification," pp. 750–761 in *Proc. SPIE Wavelet Applications*, vol. 2242, H. Szu (ed.), SPIE Pub., Bellingham, WA.

[56] M. Unser (1996), "A practical guide to the implementation of the wavelet transform," pp. 37–76 in *Wavelets in Medicine and Biology*, A. Aldroubi and M. Unser (eds.), CRC Press, New York.

[57] M. Unser, A. Aldroubi (1993), "B-spline signal processing I: Theory and II: Efficient design and application," *IEEE Trans. Signal Process.*, vol. 41, pp. 821–848.

[58] P. Vaidyanathan (1993), *Multirate Systems and Filter Banks*, Prentice–Hall, Englewood Cliffs, NJ.

[59] M. Vetterli, J. Kovačević (1995), *Wavelets and Subband Coding*, Prentice–Hall, Englewood Cliffs, NJ.

[60] M. Vrhel, C. Lee, M. Unser (1995), "Fast continuous wavelet transform," pp. 1165–1168 in *Proc. ICASSP*, vol. 2, IEEE Computer Society Press, Piscataway, NJ.

[61] V. Wickerhauser (1992), "Acoustic signal compression with wave packets," pp. 679–700 in *Wavelets: a Tutorial in Theory and Applications*, C. Chui (ed.), Academic Press, Tokyo.

[62] W. Yip, K. Leung, K. Wong (1995), "Pitch detection of speech signals in noisy environment by wavelet," pp. 604–614 in *Proc. SPIE Wavelet Applications II*, H. Szu (ed.), SPIE Pub., Bellingham, WA.

5. Wavelet Analysis of Atmospheric Wind, Turbulent Fluid, and Seismic Acceleration Data

Michio Yamada* and Fumio Sasaki†

Abstract. This chapter presents new wavelet-based techniques developed for and applied to the study of atmospheric wind, turbulent fluid, and seismic acceleration data. In our first study, we use a new wavelet-based time-frequency analysis method to separate two distinct components of atmospheric wind velocity data with energy spectra proportional to ω^{-1} and $\omega^{-5/3}$, where ω is the angular frequency. In our second study, orthonormal wavelet expansions are used to analyze turbulent velocity data from numerical simulations of the Navier–Stokes equation. More specifically, we developed a method for classifying wavelet transform coefficients from two distinct components of turbulent wind data. In our third study, we propose a method to construct biorthogonal wavelets which block diagonalize a class of integral operators and give precise conditions for the applicability of our method. Our method is applied to correct noisy seismic acceleration data.

Key words. wavelets, noise reduction, Navier–Stokes equation, time-frequency analysis, operators, seismic acceleration

5.1. Introduction

In this chapter we present new wavelet-based techniques to study atmospheric wind, turbulent fluid, and seismic acceleration data. To analyze wind data, we devise a method for classifying wavelet expansion coefficients which can be used to identify and extract two distinct components which are separated from each other in the time-frequency plane. In addition, we determine the energy and time-interval ratio of the events. To

*Graduate School of Mathematical Sciences, University of Tokyo, 3-8-1 Komaba, Meguro-ku,Tokyo 153-0041 Japan (yamada@ms.u-tokyo.ac.jp).

†Intelligent Systems Department, Kajima Corporation, 2-7 Motoakasaka 1-chome, Minato-ku, Tokyo 107-0051 Japan (fsasaki@ipc.kajima.co.jp).

analyze fluid turbulence, we examine the statistical distribution of wavelet expansion coefficients to determine the spatial distribution of Fourier coefficients. To correct seismic acceleration data, we construct a biorthogonal wavelet adapted to an integral operator, then use it, together with the Lagrange multiplier method. In the next three sections of this chapter, we describe each of the studies in detail. The noise removal techniques we develop in this chapter are more involved than that developed by Sakakibara in Chapter 2. His simpler method should be used when appropriate. Some other works on wavelets and partial differential equations are given in [3]. A useful reference for readers interested in the basics of integral equations which appear in this chapter is [19].

5.2. Extraction of Events from Time-Series

In this section, we present a method for classifying wavelet expansion coefficients of two distinct components of turbulent wind data which are separated from each other in the time-frequency plane, i.e., one is associated with frequency components localized around a particular time and the other with a different frequency and time. We use the method to determine the energy and time-interval ratio of the associated events.

When the motion of a fluid is in a state of fully developed turbulence, the time-series of the velocity measured at a spatially fixed point has an energy spectrum proportional to $\omega^{-5/3}$, where ω is the angular frequency ($2\pi \cdot$ frequency). The energy spectrum (known as the Kolmogorov spectrum) can be observed in atmospheric wind data when it is measured in free space without buoyancy effects; however, it cannot be observed in the atmospheric boundary layer (the lowest layer of the atmosphere) when it is strongly influenced by the presence of the terrestrial surface. In the boundary layer, fluid motion has strong shearing effects, and the energy spectrum is deformed. The spectral form of the energy in the boundary layer is not of Kolmogorov type. It is proportional to ω^{-1}.

Figure 5.1 is a plot of the ω^{-1} spectrum of atmospheric high wind data on a high bridge over a valley open to the sea, where the wind was blowing inland [6], [22]. We expect the spectrum to be of ω^{-1} form when the difference between the mean wind velocities at horizontally separated points on the bridge is large, i.e., the horizontal wind shear is large. Horizontal shear is produced when the wind blows along cliffs and across the bridge. An $\omega^{-5/3}$ spectrum is observed when the horizontal shear is small.

The spectral form of most of the wind data on the bridge is neither of Kolmogorov type nor proportional to ω^{-1}. It lies somewhere between these two extreme cases (Figure 5.2). If we assume that the spectral curve is a

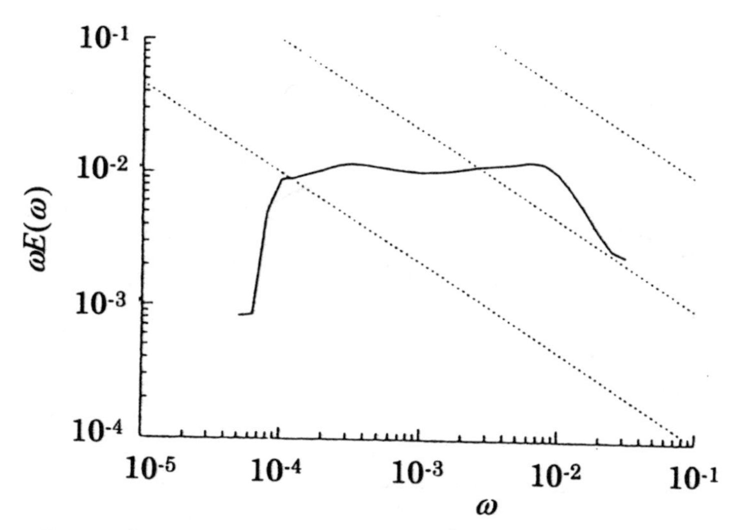

Figure 5.1. ω^{-1} *spectrum for atmospheric high wind data. The dotted line with slope* $-2/3$ *corresponds to* $E(\omega) \sim \omega^{-5/3}$.

mixture of the two cases and try to classify the time intervals corresponding to $\omega^{-5/3}$ and ω^{-1}, we are not successful. Our failure is consistent with the observation that statistical properties of the time-series appear to be almost uniform with respect to time. The mixture of $\omega^{-5/3}$ and ω^{-1} intervals depends on the scale j. Different intervals belong to the $\omega^{-5/3}$ event for different values of j. Separation is very difficult using short-time Fourier transform methods but is straightforward using wavelets.

Suppose that the time-series data is a mixture of two events which are of order $\omega^{-5/3}$ and ω^{-1} and that the pure events can be observed in the time-frequency domain but not on the time- or frequency-axes (Figure 5.3). If our assumptions are correct, then the mixture can be decomposed into two separate events using wavelet transform analysis. Wavelets

$$\psi_{j,k}(t) = 2^{j/2}\ \psi(2^j t - k); \qquad j, k \in \mathbf{Z},$$

have a time-frequency component localized near $t \sim k/2^j$ and $\omega \sim 2^j$, where $\psi(t)$ denotes a mother wavelet for an orthonormal family of functions (see section 4.1 in Chapter 4 of this book for basics on wavelet transforms and time-frequency analysis). We classify the wavelet expansion coefficients into two sets corresponding to the $\omega^{-5/3}$ and ω^{-1} events.

Our studies use orthonormal Meyer wavelets (shown in Figure 5.4)

$$\psi(t) = \frac{1}{2\pi} \int_{-\infty}^{\infty} \exp(i\omega t)\ \hat{\psi}(\omega)\ d\omega,$$

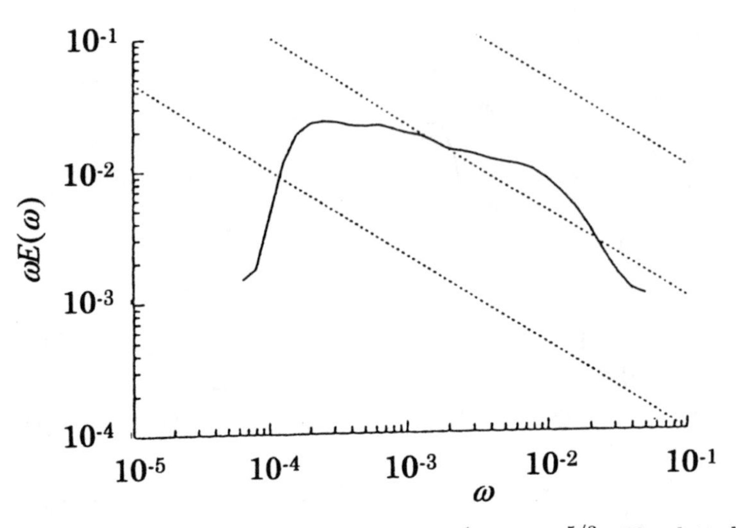

Figure 5.2. *High wind data when $E(\omega) \sim \omega^{-1}$ and $\omega^{-5/3}$. The dotted line has slope $-2/3$.*

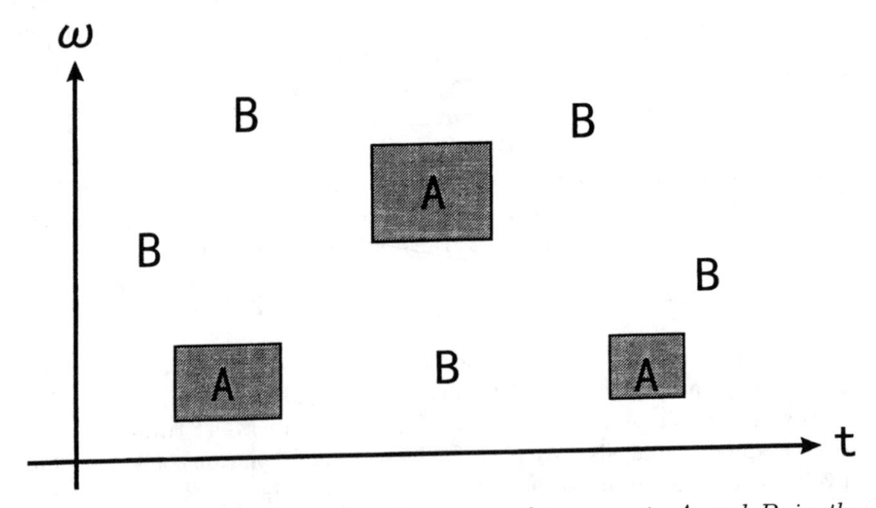

Figure 5.3. *Schematic figure of mixture of two events A and B in the time-frequency domain.*

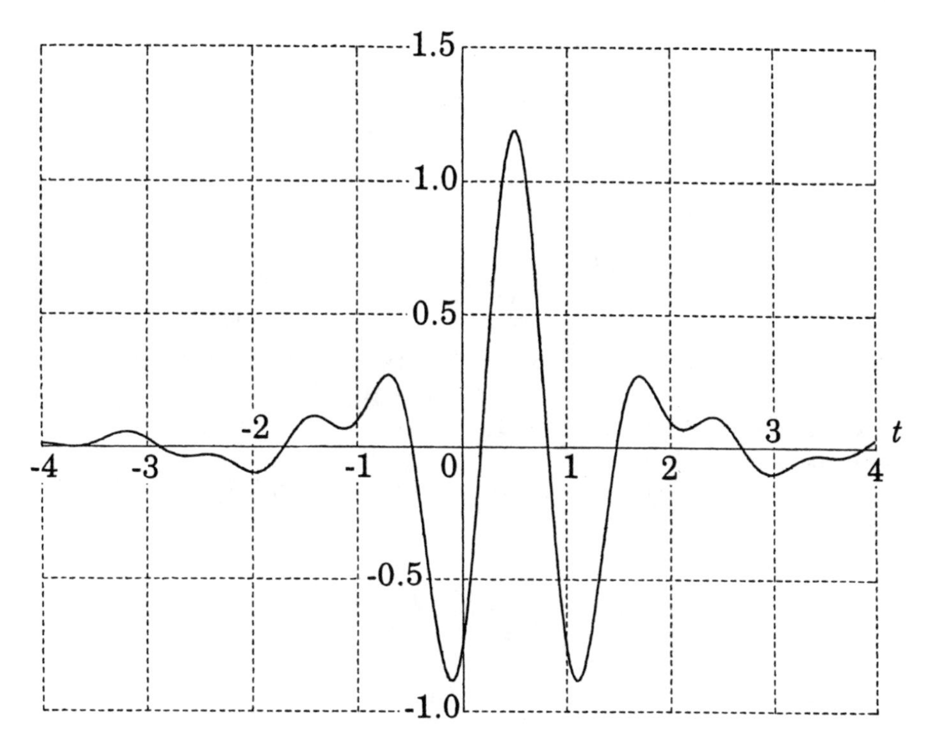

Figure 5.4. *Meyer wavelet.*

which have Fourier transform

$$\hat{\psi}(\omega) = \exp(-i\omega/2)\,\sqrt{(\hat{\phi}(\omega/2))^2 \;-\; (\hat{\phi}(\omega))^2}.$$

Note that the Fourier transform of the scaling function $\hat{\phi}(\omega)$ is not unique; it is defined as a real, infinitely differentiable function, monotonically decreasing for $\omega \geq 0$, which satisfies

1. $\hat{\phi}(\omega) \geq 0$, $\hat{\phi}(\omega) = \hat{\phi}(-\omega)$,

2. $\hat{\phi}(\omega) = \begin{cases} 1 & \text{for} \quad |\omega| \leq 2\pi/3, \\ 0 & \text{for} \quad |\omega| \geq 4\pi/3, \end{cases}$

3. $(\hat{\phi}(\omega))^2 + (\hat{\phi}(\omega - 2\pi))^2 = 1 \quad \text{for} \quad 2\pi/3 \leq \omega \leq 4\pi/3.$

We set the scaling function to be

$$\hat{\phi}(\omega) = \sqrt{g(\omega)\; g(-\omega)},$$

where

$$g(\omega) = \frac{h\ ((4\pi/3) - \omega)}{h\ (\omega - (2\pi/3))\ +\ h\ ((4\pi/3) - \omega)}$$

and

$$h(\omega) = \begin{cases} \exp(-1/\omega^2) & \text{for} \quad \omega\ >\ 0, \\ 0 & \text{for} \quad \omega\ \leq\ 0. \end{cases}$$

The Meyer wavelet is useful for our studies because it is analytic, symmetric, and its Fourier transform is compactly supported. Although the wavelet does not decay exponentially fast as $|x| \to \infty$, it decays faster than any polynomial, which is sufficient for our purposes.

Let $f(t)$ denote high wind data sampled at the rate of 1 KHz, with Meyer wavelet expansion

$$f(t) = \sum_{j,k=-\infty}^{\infty} \alpha_{j,k}\ \psi_{j,k}(t),$$

and associated energy

$$\begin{aligned} E_j \ &\equiv \ \sum_{k=-\infty}^{\infty}\ |\ \alpha_{j,k}\ |^2 \\ &= \ \frac{1}{2\pi}\int_{-\infty}^{\infty} |\ \hat{\psi}(\omega/2\pi)\ |^2 |\ \hat{f}(\omega)\ |^2\ d\omega \\ &\quad + \frac{2^{j+1}}{\pi}\mathrm{Re}\left(\int_{2\pi/3}^{4\pi/3} \overline{\hat{\psi}(\omega)}\hat{\psi}(\omega - 2\pi)\hat{f}(2^j\omega)\overline{\hat{f}(2^j(\omega - 2\pi))}\,d\omega\right) \\ &\quad + \frac{2^{j+1}}{\pi}\mathrm{Re}\left(\int_{4\pi/3}^{8\pi/3} \overline{\hat{\psi}(\omega)}\hat{\psi}(\omega - 4\pi)\hat{f}(2^j\omega)\overline{\hat{f}(2^j(\omega - 4\pi))}\,d\omega\right). \end{aligned}$$

If the Fourier components of $f(t)$ of different frequencies are not correlated, we can expect the last two integrals to vanish. If we assume that the time-series has the ergodic property, then

$$E_j \ = \ \frac{1}{2\pi}\int_{-\infty}^{\infty} |\ \hat{\psi}(\omega/2\pi)\ |^2\ |\ \hat{f}(\omega)\ |^2\ d\omega.$$

E_j is known as the *wavelet frequency spectrum.* When the Fourier spectrum is proportional to ω^{-p}, that is, $E(\omega) \sim \omega^{-p}$, the wavelet frequency spectrum has the scaling property $E_j \sim 2^{-(p-1)j}$, a property which is useful for identifying events. If the power-form of the energy spectrum exists, that is, if the energy spectrum can be expressed as a single-term polynomial in ω, then it can also be found in the wavelet frequency spectrum.

Since the Fourier transform of the Meyer wavelet is compactly supported, we do not need to compute the Fourier components of the data in very low and very high frequency regions, where the Fourier spectrum of $f(t)$ no longer takes the power-law form. Meyer wavelets are useful for observing the correspondence between the scaling behaviors of $E(\omega)$ and E_j.

When we use Meyer wavelets to study wind data, the mixture of $\omega^{-5/3}$ and ω^{-1} events in the time-frequency domain yields two distinct sets of wavelet expansion coefficients A and B, corresponding to the two separate events

$$\begin{aligned} \alpha_{j,k} \in A \quad &\text{if} \quad |\, \alpha_{j,k} \,|^2 \;\geq\; C\, E^j_{mean}, \\ \alpha_{j,k} \in B \quad &\text{if} \quad |\, \alpha_{j,k} \,|^2 \;\leq\; C\, E^j_{mean}, \end{aligned}$$

where

$$E^j_{mean} = \sqrt{\langle\, |\, \alpha_{j,k} \,|^2 \,\rangle_k},$$

and the bracket $\langle\cdot\rangle_k$ represents the average over k. Set A consists of the wavelet coefficients corresponding to the ω^{-1} event, and set B corresponds to the $\omega^{-5/3}$ event. C is a constant.

To show that sets A and B correspond to different events, we construct their wavelet frequency spectra

$$\begin{aligned} E^A_j &= \frac{1}{\#A} \sum_{k|\ (j,k)\in A} |\, \alpha_{j,k} \,|^2, \\ E^B_j &= \frac{1}{\#B} \sum_{k|\ (j,k)\in B} |\, \alpha_{j,k} \,|^2, \end{aligned}$$

where $\#A$ and $\#B$ are the number of wavelet expansion coefficients belonging to A and B. The wavelet frequency spectrum E^A_j is the wavelet frequency spectrum E_j averaged only on set A, and E^B_j is defined analogously for set B. The sets A and B consist of small pieces of intervals.

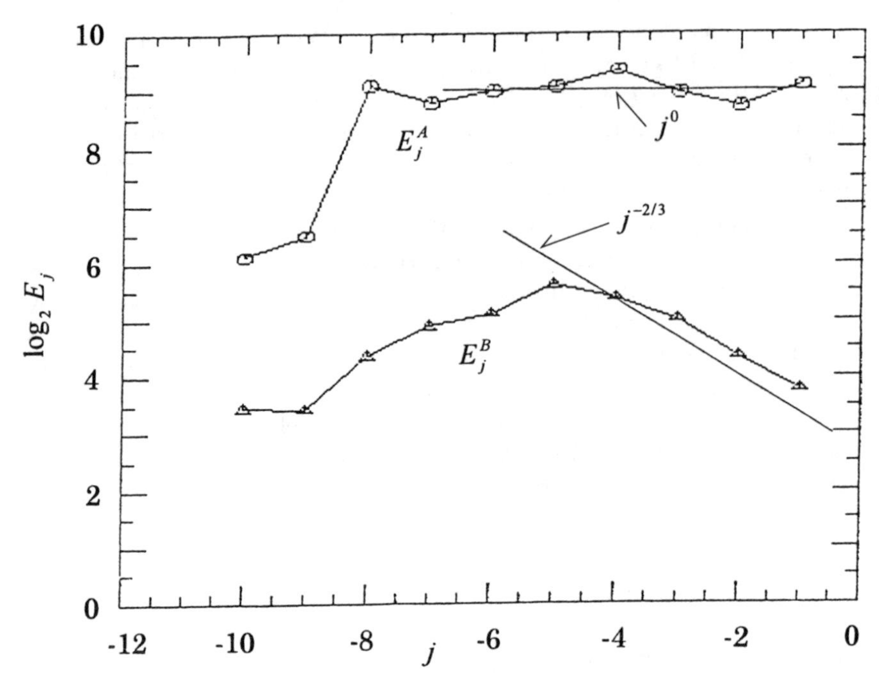

Figure 5.5. *Wavelet frequency spectra E_j^A and E_j^B.*

Data from the two events are so intertwined that they are inseparable by traditional methods of Fourier analysis.

Figure 5.5 shows the wavelet frequency spectra E_j^A and E_j^B. In the high frequency region, there is a clear separation of steepness of these spectra. The one which is almost horizontal corresponds to the ω^{-1} event, and the steeper one corresponds to the $\omega^{-5/3}$ event. Our results show that the intermediate steepness of the energy spectrum is actually a mixture of two distinct, pure events having ω^{-1} and $\omega^{-5/3}$ spectra. The time and energy ratios from thc two events are shown in Figures 5.6 and 5.7. Although the ω^{-1} event occurs 20% of the time, it has more than 50% of the energy.

To summarize, the information we obtained through wavelet analysis of wind data provided valuable scientific insight. We were able to determine the occurrence of two distinct events, their durations, and the associated energies. Further in-depth analysis of the separated "*pure*" event from the individual ω^{-1} and $\omega^{-5/3}$ spectra will likely yield more information. The wavelet-based signal separation technique, which we used, can be effectively applied in many other contexts. In particular, noise from a contaminated signal can be identified and removed [23]–[25].

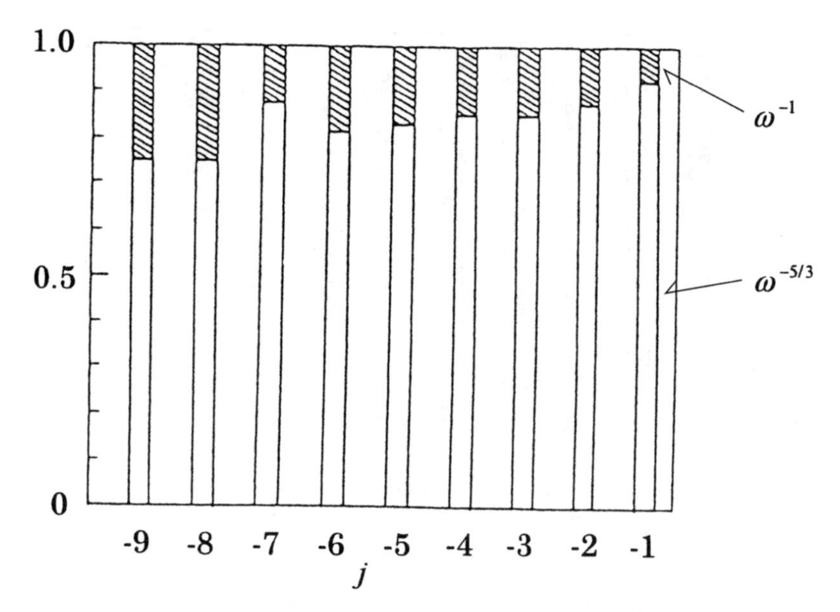

Figure 5.6. *Time ratio of events ω^{-1} and $\omega^{-5/3}$ for each j.*

5.3. Spatial Distribution of Fourier Components

In this section, we present results from wavelet analysis of turbulent velocity data from numerical simulations of the Navier–Stokes equation. More specifically, we develop a method for classifying wavelet transform coefficients from two distinct components of turbulent wind data.

Turbulent, uniform, isotropic fluids have universal properties which appear in small-scale motions of the fluid. Some statistical properties of the velocity field can be determined from the probability distribution of velocity increments

$$\Delta u(r) \equiv u(x+r) - u(x),$$

where $u(x)$ denotes the x-component of the velocity. The statistical properties of $\Delta u(r)$ are independent of the direction of the x-axis because of the isotropy and uniformity of turbulence. From the Navier–Stokes equation, Kolmogorov proved the scaling relation

$$\langle(\Delta u(r))^3\rangle \sim r^1,$$

for r sufficiently large that the viscosity of the fluid does not play a significant role in statistical properties. Here $\langle\cdot\rangle$ denotes the spatial or temporal

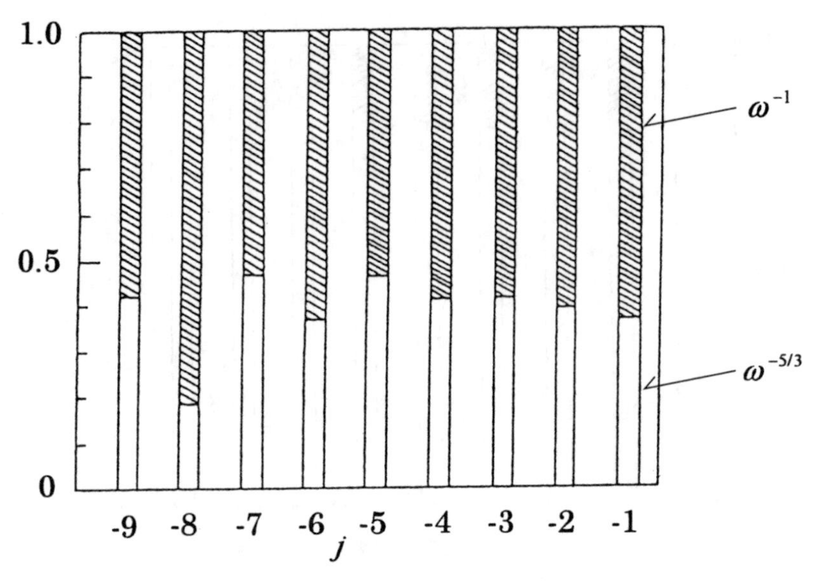

Figure 5.7. *Energy ratio of events ω^{-1} and $\omega^{-5/3}$ for each j.*

mean of the quantity. If the probability density distribution function of $\Delta u(r)$, with unit variance, is independent of r, then

$$\langle (\Delta u(r))^p \rangle \sim r^{p/3}; \qquad p \in Z.$$

This predicted behavior has been observed in numerical simulations only for small p. Observations deviate from the prediction for larger p. The functional form of the probability density distribution function $P_r(x)$ varies with r, that is, $P_r(x)$ is close to Gaussian for large r but deviates from it as r becomes smaller.

When the slope of the energy spectrum of $u(x)$ satisfies certain conditions, the distribution of the velocity increment $\Delta u(r)$ indicates the spatial distribution of Fourier components (FCs) with wavenumber $k \sim (1/r)$. For example, $(\Delta u(r))^2$ is determined exclusively by components with $k \sim (1/r)$ only if the energy spectrum $E(k)$ has slope between k^{-1} and k^{-3}. Moreover, the set of all velocity increments $\Delta u(r)$ cannot be taken as independent quantities reflecting the spatial distribution of the FCs since the number of FCs with wavenumber $k \sim (1/r)$ is less than the total number of points in real space.

In our study, we use Meyer wavelets, which have compact support in Fourier space. The wavelet expansion coefficients reflect the spatial distribution of FCs in a limited domain in Fourier space so that there is no contamination by larger- or smaller-wavenumber components. Furthermore, these coefficients are clearly independent of each other due to the orthogonality of Meyer wavelets. We determined the probability density function (PDF) $P(x)$ of the wavelet expansion coefficients $\alpha_{j,k}$ for each j of $u(x)$, where

$$u(x) = \sum_{j,k} \alpha_{j,k} \; \psi_{j,k}(x)$$

[26]. The velocity data $u(x)$ was produced from numerical simulations using the Navier–Stokes equation with symmetric initial and boundary conditions, a Fourier spectral, and a fourth-order Runge–Kutta method. The energy spectrum of u is given in Figure 5.6. The effective number of modes is 128^8. The aliasing error was removed using the 2/3 rule. The kinetic viscosity was 0.0005, which yielded a Taylor microscale Reynolds number $R_\lambda \sim 180$.

Figure 5.8 shows the PDF for several values of j, where a larger j corresponds to a higher wavenumber band. The PDF is obtained by constructing a histogram of $\alpha_{j,k}$ over k for each value of j. The solid curve represents a normal Gaussian distribution. The PDF is near Gaussian for small j, deviates noticeably for $j \geq 6$, and does not decay exponentially. We set $x^{1/2}$ and $\log P(x)$ to be the x- and y-axes and observe that the PDFs for $j = 8$ and 9 have a stretched exponential tail, that is,

$$P(x) \;\propto\; \exp(-\text{constant} \cdot x^{1/2}),$$

(see Figure 5.9) which is consistent with experimental results [11].

In our simulations, the viscosity of the fluid influences the flow field for wavenumbers $j \geq 8$. The PDF in this range reflects the effect of viscosity on the spatial distribution of Fourier components. In our graphs, $P(x)$ appears to be of stretched exponential form for $x > x_{cr}$, with larger x_{cr} for larger j. For $x < x_{cr}$, the slope of the PDF is steeper than the stretched exponential form. Figure 5.10 shows the PDFs on a log-log scale. The solid curves represent the standard Gaussian distribution. For $j = 8$ and 9, the PDF is approximated well by the power form of $x^{-1.6}$ for small x, but deviates for larger x. For large x, the PDF looks like a stretched exponential. This observation suggests that if a larger value of j is available from better (e.g., higher Reynolds number) simulation data, the stretched exponential form will be driven out to the regions where X is very large and will be replaced by a power form in most regions of x [26].

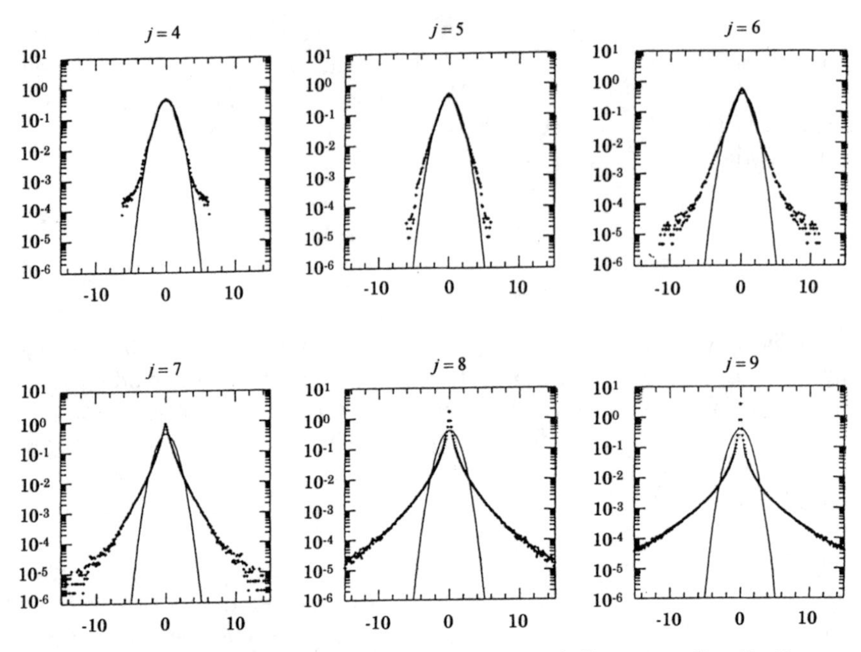

Figure 5.8. *The PDFs of $\alpha_{j,k}$ vs. the normal Gaussian distribution.*

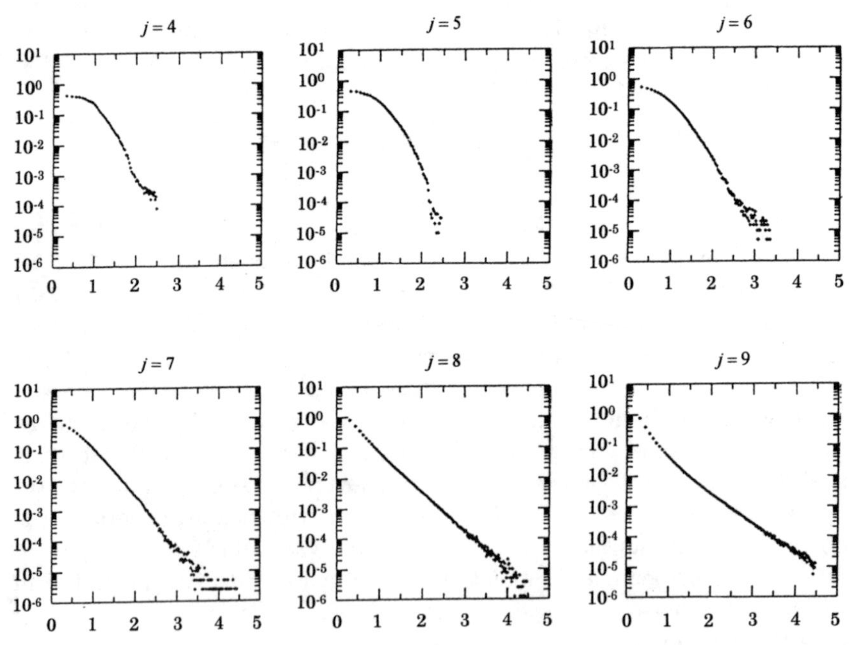

Figure 5.9. *The PDFs of $\alpha_{j,k}$, with ordinate $x^{1/2}$.*

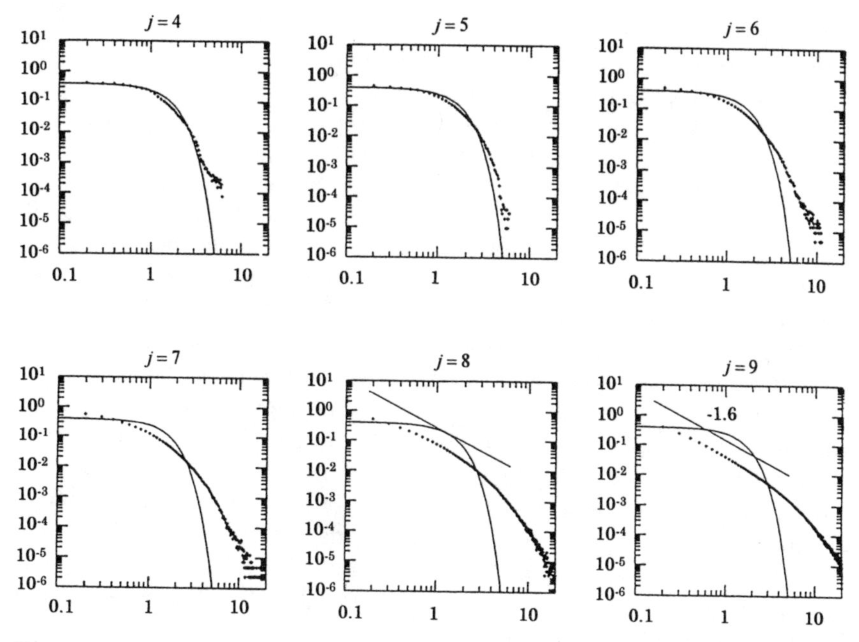

Figure 5.10. *Log-log plots of the PDFs of* $\alpha_{j,k}$. *The lines have slope* $x^{-1.6}$.

5.4. Correction of Seismic Data

In this section we construct biorthogonal wavelets adapted to a class of integral operators, then propose a new $O(n)$ method for solving differential equations with operators of odd order (d^{2l+1}/dx^{2l+1}). The inverse of the differential operator is regarded as a sum of two integral operators, each of which can be block diagonalized using a biorthogonal wavelet representation. We develop a new data correction method in the time-frequency domain using our biorthogonal wavelet and apply the method to clean noisy seismic acceleration data [17].

Base line correction and noise removal in high and low frequency bands of observational seismic acceleration data are required before the measurements can be used to compute the corresponding velocity and displacement by numerical integration. Fourier transform–based methods can be used for correction by eliminating the contaminated frequency band, if noise is present only in high or low frequency regions, throughout the entire period when measurements are recorded. However, if high frequency noise is only present for a limited period during the measurement interval, data correction should be performed in the time-frequency domain, using a more suitable approach, such as wavelet analysis.

Currently, Wavelet–Galerkin methods are the most widely used wavelet-based numerical methods for solving differential equations [1], [5], [12], [13], [20]. In these methods, the solution is expressed as a wavelet expansion, the coefficients of which must be determined. Qian and Weiss assume periodic boundary conditions and use fast Fourier transforms [16]. Dahlke and Weinreich developed a new wavelet such that the matrix representation of an even-order differential operator (d^{2l}/dx^{2l}) is block diagonal [2]. Williams and Amaratunga proved that the Dahlke–Weinreich representation is almost diagonal and that the problem is of $O(n)$, for an n-by-n matrix [21]. The wavelets constructed by Dahlke and Weinreich are biorthogonal. They cannot be orthonormal, because orthonormality cannot be maintained when making the matrix representation of the differential operator block diagonal. Jameson constructed differentiation matrices for Daubechies' and spline-based wavelet bases [7], [9], [10]. For Daubechies' wavelets with M vanishing moments, the matrix is accurate to order $2M$ for periodic boundary conditions. For an nth-order spline basis, the matrix has accuracy of order $2n+2$ for periodic boundary conditions. These orders of accuracy cannot be guaranteed when boundary conditions are nonperiodic. Jameson also proposed a wavelet-optimized finite difference method which does not suffer from difficulties with nonlinear and boundary terms since all calculations are performed in physical space [8].

5.4.1. Notation and Preliminaries

In this subsection, we establish our notation and review concepts which are used in the remainder of this paper. We begin with a review of biorthogonal wavelets. (Readers may want to compare the concepts with those for multiresolution analyses (MRA) using orthonormal wavelets, given in section 1.2 of Chapter 1.) Let

$$\cdots \subset V_{-2} \subset V_{-1} \subset V_0 \subset V_1 \subset V_2 \subset \cdots$$

$$\cdots \subset \tilde{V}_{-2} \subset \tilde{V}_{-1} \subset \tilde{V}_0 \subset \tilde{V}_1 \subset \tilde{V}_2 \subset \cdots$$

be two nested subsequences of closed subspaces of $\mathbf{L}^2(\mathbf{R})$ associated with two multiresolution ladders with primary and dual scaling functions ϕ and $\tilde{\phi}$ which generate the bases of the spaces V_j and $\tilde{V}_j$

$$\{\phi_{j,k}(t) \;:\; \phi_{j,k}(t) = 2^{j/2}\,\phi(2^j t - k); \quad j,k \in \mathbf{Z}\},$$
$$\{\tilde{\phi}_{j,k}(t) \;:\; \tilde{\phi}_{j,k}(t) = 2^{j/2}\,\tilde{\phi}(2^j t - k); \quad j,k \in \mathbf{Z}\}.$$

Let W_j be the complement of V_j in V_{j+1} and $\tilde{W}_j$ the complement of $\tilde{V}_j$ in $\tilde{V}_{j+1}$.

$$V_{j+1} = V_j \oplus W_j, \quad \tilde{V}_{j+1} = \tilde{V}_j \oplus \tilde{W}_j, \tag{5.1}$$

and

$$W_j \perp \tilde{V}_j, \quad \tilde{W}_j \perp V_j.$$

W_j and $\tilde{W}_j$ are not necessarily orthogonal complements of V_j and $\tilde{V}_j$. They are generated by integer translations of the primary and dual wavelets $\psi(t) \in W_0$ and $\tilde{\psi}(t) \in \tilde{W}_0$, where

$$\{\psi_{j,k}(t) \ : \ \psi_{j,k}(t) \ = \ 2^{j/2} \ \psi(2^j t - k); \quad j,k \in \mathbf{Z}\},$$
$$\{\tilde{\psi}_{j,k}(t) \ : \ \tilde{\psi}_{j,k}(t) \ = \ 2^{j/2} \ \tilde{\psi}(2^j t - k); \quad j,k \in \mathbf{Z}\}.$$

Conditions for biorthogonality are given by

$$\langle \phi(\cdot), \ \tilde{\phi}(\cdot - k)\rangle = \delta_{0,k}, \qquad \langle \psi_{j,k}, \ \tilde{\psi}_{j',k'}\rangle = \delta_{j,j'} \cdot \delta_{k,k'} \tag{5.2}$$

$$\langle \psi(\cdot), \ \tilde{\phi}(\cdot - k)\rangle = 0, \qquad \langle \tilde{\psi}(\cdot), \ \phi(\cdot - k)\rangle = 0. \tag{5.3}$$

The filter bank coefficients h, $\tilde{h}$, g, and $\tilde{g}$ are used to express dilation relations between scaling functions and wavelets

$$\phi(t) = \sum_k \sqrt{2} \ h_k \ \phi(2t-k), \qquad \psi(t) = \sum_k \sqrt{2} \ g_k \ \phi(2t-k),$$
$$\tilde{\phi}(t) = \sum_k \sqrt{2} \ \tilde{h}_k \ \tilde{\phi}(2t-k), \qquad \tilde{\psi}(t) = \sum_k \sqrt{2} \ \tilde{g}_k \ \tilde{\phi}(2t-k),$$

and the scaling relations

$$\phi_{j+1,k} \ = \ \sum_l \ \overline{\tilde{h}}_{k-2l} \ \phi_{j,l} \ + \ \sum_l \ \overline{\tilde{g}}_{k-2l} \ \psi_{j,l},$$
$$\tilde{\phi}_{j+1,k} \ = \ \sum_l \ \overline{h}_{k-2l} \ \tilde{\phi}_{j,l} + \sum_l \ \overline{g}_{k-2l} \ \tilde{\psi}_{j,l}.$$

The Fourier transform of the wavelets and scaling functions are

$$\hat{\phi}(2\omega) = m_0(\omega) \ \hat{\phi}(\omega), \qquad \hat{\tilde{\phi}}(2\omega) = \tilde{m}_0(\omega) \ \hat{\tilde{\phi}}(\omega), \tag{5.4}$$

$$\hat{\psi}(2\omega) = m_1(\omega) \ \hat{\phi}(\omega), \qquad \hat{\tilde{\psi}}(2\omega) = \tilde{m}_1(\omega) \ \hat{\tilde{\phi}}(\omega), \tag{5.5}$$

where ˆ denotes the Fourier transform of a function and

$$m_0(\omega) = \sum_k \ (h_k/\sqrt{2}) \ e^{-ik\omega}, \qquad \tilde{m}_0(\omega) = \sum_k \ (\tilde{h}_k/\sqrt{2}) \ e^{-ik\omega}, \tag{5.6}$$

$$m_1(\omega) = \sum_k \ (g_k/\sqrt{2}) \ e^{-ik\omega}, \qquad \tilde{m}_1(\omega) = \sum_k \ (\tilde{g}_k/\sqrt{2}) \ e^{-ik\omega}. \tag{5.7}$$

$m_0(\omega)$ and $m_1(\omega)$ are known as the primary symbols associated with h_k and g_k, and $\tilde{m}_0(\omega)$ and $\tilde{m}_1(\omega)$ the dual symbols associated with $\tilde{h}_k$ and $\tilde{g}_k$. The biorthogonal conditions of equations (5.2)–(5.3) are represented in terms of these symbols as

$$m_0(\omega)\,\overline{\tilde{m}_0(\omega)} + m_0(\omega+\pi)\,\overline{\tilde{m}_0(\omega+\pi)} = 1, \tag{5.8}$$

$$m_1(\omega)\,\overline{\tilde{m}_1(\omega)} + m_1(\omega+\pi)\,\overline{\tilde{m}_1(\omega+\pi)} = 1,$$

$$m_1(\omega)\,\overline{\tilde{m}_0(\omega)} + m_1(\omega+\pi)\,\overline{\tilde{m}_0(\omega+\pi)} = 0,$$

$$\tilde{m}_1(\omega)\,\overline{m_0(\omega)} + \tilde{m}_1(\omega+\pi)\,\overline{m_0(\omega+\pi)} = 0,$$

where the overline denotes the complex conjugation. The conditions for $m_1(\omega)$ and $\tilde{m}_1(\omega)$ are satisfied if $m_1(\omega)$ and $\tilde{m}_1(\omega)$ satisfy

$$m_1(\omega) = -e^{-i\omega}\,\overline{\tilde{m}_0(\omega+\pi)}, \quad \tilde{m}_1(\omega) = -e^{-i\omega}\,\overline{m_0(\omega+\pi)}, \tag{5.9}$$

so that

$$g_k = (-1)^k\,\overline{\tilde{h}}_{1-k}, \quad \tilde{g}_k = (-1)^k\,\overline{h}_{1-k}. \tag{5.10}$$

When the coefficients $\{h_k\}$ and $\{\tilde{h}_k\}$ of the scaling functions $\phi(t)$ and $\tilde{\phi}(t)$ are given, the primary and dual wavelets can be calculated from equations (5.4), (5.5), and (5.9)–(5.10).

We construct relations between expansion coefficients similar to Mallat's transformations for orthonormal wavelets [14]. Consider the projection of $f(t) \in \mathbf{L}^2(\mathbf{R})$ onto the closed subspaces V_j and $\tilde{V}_j$.

$$\begin{aligned} \text{Proj}\,(f(t))_{V_j} &= \sum_k c_{j,k}\,\phi_{j,k}(t), \\ \text{Proj}\,(f(t))_{\tilde{V}_j} &= \sum_k \tilde{c}_{j,k}\,\tilde{\phi}_{j,k}(t), \end{aligned}$$

where $c_{j,k} = \langle f(t),\, \tilde{\phi}_{j,k}(t)\rangle$ and $\tilde{c}_{j,k} = \langle f(t),\, \phi_{j,k}(t)\rangle$ and the projections onto W_j and $\tilde{W}_j$,

$$\begin{aligned} \text{Proj}\,(f(t))_{W_j} &= \sum_k d_{j,k}\,\psi_{j,k}(t), \\ \text{Proj}\,(f(t))_{\tilde{W}_j} &= \sum_k \tilde{d}_{j,k}\,\tilde{\psi}_{j,k}(t), \end{aligned}$$

where $d_{j,k} = \langle f(t),\, \tilde{\psi}_{j,k}(t)\rangle$ and $\tilde{d}_{j,k} = \langle f(t),\, \psi_{j,k}(t)\rangle$. Calculations similar to those for orthonormal wavelets lead to the following relations between the expansion coefficients.

Decomposition:

$$c_{j-1,k} = \sum_l \overline{\tilde{h}}_{l-2k}\ c_{j,l}, \quad d_{j-1,k} = \sum_l \overline{\tilde{g}}_{l-2k}\ c_{j,l};$$

Reconstruction:

$$c_{j,l} = \sum_k h_{l-2k} c_{j-1,k} + \sum_k g_{l-2k} d_{j-1,k};$$

Decomposition:

$$\tilde{c}_{j-1,k} = \sum_l \overline{h}_{l-2k}\ \tilde{c}_{j,l}, \quad \tilde{d}_{j-1,k} = \sum_l \overline{g}_{l-2k}\ \tilde{c}_{j,l};$$

Reconstruction:

$$\tilde{c}_{j,l} = \sum_k \tilde{h}_{l-2k}\ \tilde{c}_{j-1,k} + \sum_k \tilde{g}_{l-2k}\ \tilde{d}_{j-1,k}.$$

Consider the equation

$$A\ u(t) = f(t), \tag{5.11}$$

with periodic conditions, where A is a linear operator, $f(t) \in \mathbf{L}^2(\mathbf{R}/\mathbf{Z})$ a given function, and $u(t) \in \mathbf{L}^2(\mathbf{R}/\mathbf{Z})$ is an unknown function. We will construct a numerical algorithm to solve this equation using biorthogonal wavelets.

Let $u_j(t)$ and $f_j(t)$ be Galerkin approximations of $u(t)$ and $f(t)$ using the jth level scaling function and dual

$$u_j(t) = \sum_k c_{j,k}\ \phi_{j,k}(t), \quad f_j(t) = \sum_k \tilde{s}_{j,k}\ \tilde{\phi}_{j,k}(t),$$

where $c_{j,k} = \langle u_j(t),\ \tilde{\phi}_{j,k}(t)\rangle$ and $\tilde{s}_{j,k} = \langle f_j(t),\ \phi_{j,k}(t)\rangle$. We substitute the expressions $u_j(t)$ and $f_j(t)$ into (5.11) and compute the inner product with $\{\phi_{j,m}\}$ to obtain the Wavelet–Galerkin representation

$$\sum_{k=0}^{2^j-1} c_{j,k} \left\langle \sum_{l=-\infty}^{\infty} A\ \phi_{j,k+l\cdot 2^j},\ \phi_{j,m} \right\rangle = \tilde{s}_{j,m},$$

which we rewrite in a matrix form as

$$\mathbf{K_j}\ \mathbf{c_j} = \tilde{\mathbf{s}}_\mathbf{j}. \tag{5.12}$$

Here $\mathbf{c_j} = \{c_{j,0},\ c_{j,1},\ \ldots,\ c_{j,2^j-1}\}^T$, $\mathbf{\tilde{s}_j} = \{\tilde{s}_{j,0},\ \tilde{s}_{j,1},\ \ldots,\ \tilde{s}_{j,2^j-1}\}^T$, and $\mathbf{K_j}$ is a $2^j \times 2^j$ matrix with (k,m)th element

$$(\mathbf{K_j})_{k,m} = \left\langle \sum_{l=-\infty}^{\infty} A\ \phi_{j,k+l\cdot 2^j},\ \phi_{j,m} \right\rangle.$$

To move equation (5.12) from the jth resolution scale to the $(j-1)$th scale, we use the direct sum relations from equation (5.1) to decompose $u_j(t)$ and $f_j(t)$ as

$$u_j(t) = u_{j-1}(t) + v_{j-1}(t), \tag{5.13}$$

$$f_j(t) = f_{j-1}(t) + g_{j-1}(t), \tag{5.14}$$

where

$$v_{j-1}(t) = \sum_k d_{j-1,k}\ \psi_{j-1,k}(t),$$

$$g_{j-1}(t) = \sum_k \tilde{t}_{j-1,k}\ \tilde{\psi}_{j-1,k}(t).$$

Substitution of equations (5.13)–(5.14) into (5.11) leads to the Wavelet–Galerkin representation in the $(j-1)$th resolution scale.

$$\begin{bmatrix} \mathbf{K_{j-1}} & \mathbf{\Lambda_{j-1}} \\ \mathbf{\Delta_{j-1}} & \mathbf{\Gamma_{j-1}} \end{bmatrix} \begin{pmatrix} \mathbf{c_{j-1}} \\ \mathbf{d_{j-1}} \end{pmatrix} = \begin{pmatrix} \mathbf{\tilde{s}_{j-1}} \\ \mathbf{\tilde{t}_{j-1}} \end{pmatrix}, \tag{5.15}$$

where

$$\mathbf{d_{j-1}} = (d_{j-1,0},\ d_{j-1,1},\ \ldots,\ d_{j-1,2^{j-1}-1})^T,$$

$$\mathbf{\tilde{t}_{j-1}} = (\tilde{t}_{j-1,0},\ \tilde{t}_{j-1,1},\ \ldots,\ \tilde{t}_{j-1,2^{j-1}-1})^T,$$

$$(\mathbf{K_{j-1}})_{k,m} = \left\langle \sum_{l=-\infty}^{\infty} A\ \phi_{j-1,k+l\cdot 2^{j-1}},\ \phi_{j-1,m} \right\rangle,$$

$$(\mathbf{\Gamma_{j-1}})_{k,m} = \left\langle \sum_{l=-\infty}^{\infty} A\ \psi_{j-1,k+l\cdot 2^{j-1}},\ \psi_{j-1,m} \right\rangle,$$

$$(\mathbf{\Delta_{j-1}})_{k,m} = \left\langle \sum_{l=-\infty}^{\infty} A\ \phi_{j-1,k+l\cdot 2^{j-1}},\ \psi_{j-1,m} \right\rangle,$$

$$(\mathbf{\Lambda_{j-1}})_{k,m} = \left\langle \sum_{l=-\infty}^{\infty} A\ \psi_{j-1,k+l\cdot 2^{j-1}},\ \phi_{j-1,m} \right\rangle.$$

5.4.2. Block-Diagonal Representation

Dahlke and Weinreich [2] constructed a wavelet basis in which the representation matrix for the even-order differential operator is block diagonal, so that in equation (5.15),

$$\Delta_{\mathbf{j}} = \Lambda_{\mathbf{j}} = \mathbf{0}; \qquad \forall\, j \in \mathbf{Z}.$$

We construct biorthogonal wavelets in which the representation matrices of some specific operators are block diagonal or semi-block diagonal, i.e.,

$$\Delta_{\mathbf{j}} = \mathbf{0} \quad \text{or} \quad \Lambda_{\mathbf{j}} = \mathbf{0}; \qquad \forall\, j \in \mathbf{Z}.$$

The condition $\Delta_{\mathbf{0}} = \mathbf{0}$ is satisfied if the biorthogonal wavelet and the scaling function satisfy

$$\langle A\phi(\cdot),\ \psi(\cdot - k)\rangle = 0; \qquad \forall\, k \in \mathbf{Z}. \tag{5.16}$$

We interpret this relation in terms of the symbols $m_0(\omega)$ and $m_1(\omega)$ as follows.

$$\begin{aligned}
&\langle A\phi(t),\ \psi(t-k)\rangle \\
&= \frac{1}{2\pi}\int_{-\infty}^{\infty} \hat{A}[\omega]\ \hat{\phi}(\omega)\ \overline{\hat{\psi}(\omega - k)}\ d\omega \\
&= \frac{1}{2\pi}\int_{-\infty}^{\infty} e^{ik\omega}\ \hat{A}[\omega]\ \hat{\phi}(\omega)\ \overline{\hat{\psi}(\omega)}\ d\omega \\
&= \frac{1}{2\pi}\int_{-\infty}^{\infty} e^{ik\omega}\ \hat{A}[\omega]\ m_0\left(\frac{\omega}{2}\right)\ \hat{\phi}\left(\frac{\omega}{2}\right)\ \overline{m_1\left(\frac{\omega}{2}\right)\hat{\phi}\left(\frac{\omega}{2}\right)}\ d\omega \\
&= \frac{1}{2\pi}\sum_{n\in\mathbf{Z}}\int_{[0,2\pi]+2n\pi} e^{ik\omega}\ \hat{A}[\omega]\ m_0\left(\frac{\omega}{2}\right)\ \overline{m_1\left(\frac{\omega}{2}\right)}\ \left|\hat{\phi}\left(\frac{\omega}{2}\right)\right|^2 d\omega \\
&= \frac{1}{2\pi}\sum_{n\in\mathbf{Z}}\int_{0}^{2\pi} e^{ik\omega}\ \hat{A}[\omega + 2n\pi]\ m_0\left(\frac{\omega+2n\pi}{2}\right) \\
&\quad \times \overline{m_1\left(\frac{\omega+2n\pi}{2}\right)}\ \left|\hat{\phi}\left(\frac{\omega+2n\pi}{2}\right)\right|^2 d\omega
\end{aligned}$$

$$= \frac{1}{2\pi} \int_0^{2\pi} e^{ik\omega} \left(m_0\left(\frac{\omega}{2}\right) \overline{m_1\left(\frac{\omega}{2}\right) \sum_{n\in\mathbf{Z}} \hat{A}[\omega + 4n\pi] \left|\hat{\phi}\left(\frac{\omega}{2} + 2n\pi\right)\right|^2} \right.$$

$$\left. + m_0\left(\frac{\omega}{2}+\pi\right) \overline{m_1\left(\frac{\omega}{2}+\pi\right) \sum_{n\in\mathbf{Z}} \hat{A}[\omega+2\pi+4n\pi] \left|\hat{\phi}\left(\frac{\omega}{2}+\pi+2n\pi\right)\right|^2} \right) d\omega.$$

Let $g(\omega)$ be the associate function of operator A defined by

$$g(\omega) = \sum_{n\in\mathbf{Z}} \hat{A}[\omega + 4n\pi] \left|\hat{\phi}\left(\frac{\omega}{2} + 2n\pi\right)\right|^2.$$

Then equation (5.16) can be rewritten as

$$\int_0^{2\pi} e^{ik\omega} \left(m_0\left(\frac{\omega}{2}\right) \overline{m_1\left(\frac{\omega}{2}\right) g(\omega)} \right.$$

$$\left. + m_0\left(\frac{\omega}{2}+\pi\right) \overline{m_1\left(\frac{\omega}{2}+\pi\right) g(\omega+2\pi)} \right) d\omega = 0$$

for all $k \in \mathbf{Z}$. A necessary and sufficient condition for satisfying the above is

$$m_0(\omega)\ \tilde{m}_0(\omega+\pi)\ g(2\omega) \ - \ m_0(\omega+\pi)\ \tilde{m}_0(\omega)\ g(2\omega+2\pi) = 0. \quad (5.17)$$

From the biorthogonality conditions (5.8) and (5.17), we have the dual symbol

$$\tilde{m}_0(\omega) \ = \ \frac{m_0(\omega)\ g(2\omega)}{|m_0(\omega)|^2\ g(2\omega)\ +\ |m_0(\omega+\pi)|^2\ g(2\omega+2\pi)}. \quad (5.18)$$

Once we are given the primary symbol $m_0(\omega)$, we can construct the dual $\tilde{m}_0(\omega)$ so long as the denominator of the right-hand side of equation (5.18) does not vanish.

To satisfy $\Lambda_\mathbf{0} = \mathbf{0}$, the biorthogonal wavelet must satisfy

$$\langle A\ \psi(\cdot)\ ,\ \phi(\cdot - k)\rangle = 0,$$

a necessary and sufficient condition for which is

$$m_0(\omega)\ \tilde{m}_0(\omega+\pi)\ \overline{g(2\omega)} \ - \ m_0(\omega+\pi)\ \tilde{m}_0(\omega)\ \overline{g(2\omega+2\pi)} \ = \ 0.$$

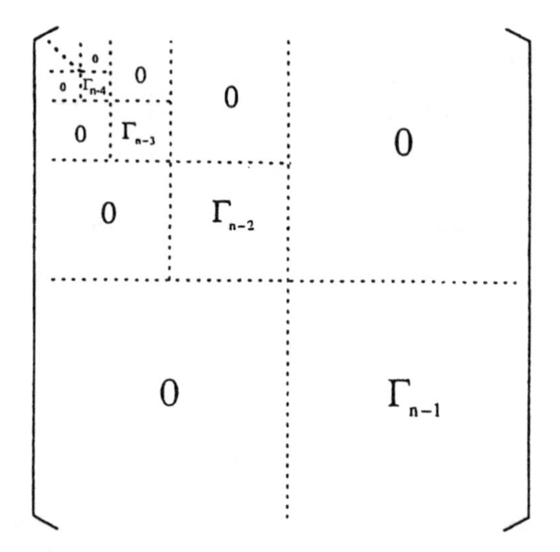

Figure 5.11. *The representation matrix.*

This condition is, in turn, satisfied if (5.17) holds and $g(\omega)$ is a real or pure imaginary function.

If we restrict ourselves to linear operators which satisfy

$$[Af(\lambda\cdot)](t) = \lambda^{\alpha}\ [Af(\cdot)](\lambda t); \quad \forall\, \lambda \in \mathbf{R}^{+},$$

where α is a fixed constant, then the representation matrix satisfies

$$\Delta_{\mathbf{j}} = 2^{j\alpha}\ \Delta_{\mathbf{0}}, \quad \Lambda_{\mathbf{j}} = 2^{j\alpha}\ \Lambda_{\mathbf{0}},$$

which means that if $\Delta_{\mathbf{0}} = \Lambda_{\mathbf{0}} = \mathbf{0}$, then $\Delta_{\mathbf{j}} = \Lambda_{\mathbf{j}} = \mathbf{0}$ for all j, that is, A is block diagonal and is of the form shown in Figure 5.11.

To summarize, given an operator A and the primary symbol $m_0(\omega)$, such that

(i) A is invariant to scale transformations,
(ii) $\tilde{m}_0(\omega)$ does not diverge, and
(iii) either $g(\omega) = \overline{g(\omega)}$ or $ig(\omega)$ is a real function.

We can construct an adaptive biorthogonal wavelet which leads to a block-diagonal representation of the operator A. When condition (iii) is not satisfied ($g(\omega)$ is neither real nor pure imaginary), $\Lambda_{\mathbf{0}}$ does not necessarily vanish and the representation matrix is semi-block-diagonal.

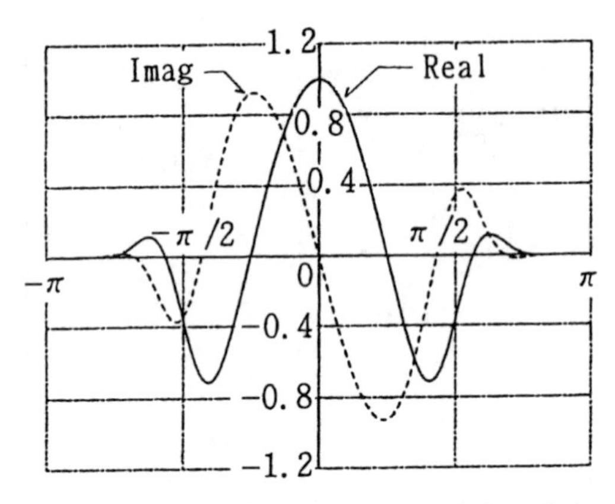

Figure 5.12. *The primary symbol $m_0(\omega)$.*

5.4.3. Adaptive Biorthogonal Wavelets and Block-Diagonal Operators

In this subsection, we explicitly construct biorthogonal scaling functions and wavelets adapted to three integral operators: the Riesz potential [18], the first derivative of the Hilbert transform [18], and the Abel transform [19]. These operators are invariant to scale transformation, and the corresponding dual symbol $\tilde{m}_0(\omega)$ is well defined. The corresponding representation matrix is block diagonal in the first two cases but is semi-block-diagonal in the third case because $g(\omega)$ is neither real nor pure imaginary. In the examples below, we use the primary symbol $m_0(\omega)$ from Daubechies' compactly supported orthonormal wavelet (N=6), shown in Figure 5.12.

$$m_0(\omega) = \left(\frac{1+e^{-i\omega}}{2}\right) m_{Daub}(\omega),$$

where $m_{Daub}(\omega)$ is the symbol of the Daubechies' wavelet [4]. $m_0(\omega)$ is a smooth function of ω. The primary scaling function constructed from $m_0(\omega)$ is compactly supported in the temporal domain.

The **Riesz potential** is defined by

$$I_\alpha[u(x)] = \frac{1}{\gamma(\alpha)} \int_{-\infty}^{\infty} \frac{u(y)\, dy}{|x-y|^{1-\alpha}}; \qquad 0 < \alpha < 1,$$

where

$$\gamma(\alpha) = \frac{\sqrt{\pi}\; 2^\alpha\; \Gamma(\alpha/2)}{\Gamma((1-\alpha)/2)},$$

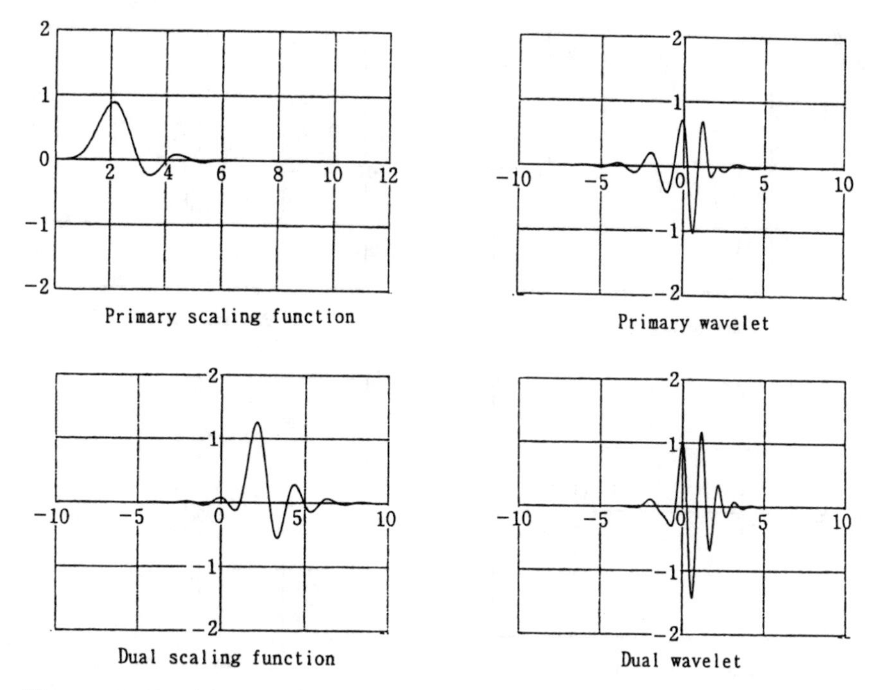

Figure 5.13. *Adaptive scaling functions and biorthogonal wavelets for the Riesz potential.*

and Γ is the Gamma function [18]. I_α is invariant to transformation in scale, i.e.,

$$I_\alpha[u(\lambda\cdot)](x) = \lambda^{-\alpha}\ I_\alpha[u(\cdot)](\lambda x),$$

and in Fourier space, the operator I_α is equivalent to multiplication by $1/|\omega|^\alpha$, i.e.,

$$\widehat{I_\alpha}[u(\cdot)](\omega) = \frac{1}{|\omega|^\alpha}\ \hat{u}(\omega).$$

The associate function $g(\omega)$ of the operator I_α is given by

$$g(\omega) = \sum_{n=-\infty}^{\infty} \frac{1}{|\omega\ +\ 4n\pi|^\alpha}\ \left|\hat{\phi}\left(\frac{\omega}{2}\ +\ 2n\pi\right)\right|^2.$$

The primary symbol $m_0(\omega)$ and the associated function $g(\omega)$ are used to determine the dual symbol $\tilde{m}_0(\omega)$. $\tilde{m}_0(\omega)$ is a well-defined, smooth function

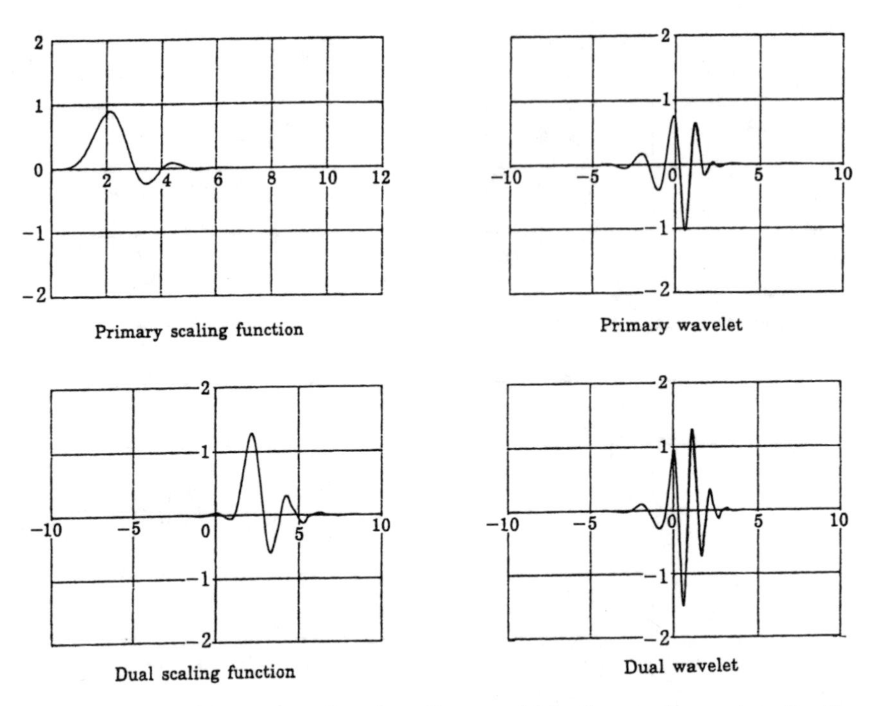

Figure 5.14. *Adaptive scaling functions and biorthogonal wavelets for the first derivative of the Hilbert transform.*

of ω, and the associate function $g(\omega)$ satisfies $g(\omega) = \overline{g(\omega)}$. The three conditions given in subsection 5.4.2, that guarantee a block-diagonal representation of A using adaptive biorthogonal wavelets, are satisfied. The adaptive scaling functions and the biorthogonal wavelets of I_α for $\alpha = 0.5$ are shown in Figure 5.13.

The **first derivative of the Hilbert transform** is defined by

$$H'[u(x)] = \frac{d}{dx}\left[\frac{1}{\pi}\ vp \int_{-\infty}^{\infty} \frac{u(y)\ dy}{y-x}\right],$$

where vp indicates the principal value of the integral [18]. H' is invariant to transformations in scale,

$$H'[u(\lambda\cdot)](x) = \lambda\ H'[u(\cdot)](\lambda x);$$

its Fourier transform is

$$\widehat{H'}[u(\cdot)](\omega) = -|\omega|\ \hat{u}(\omega),$$

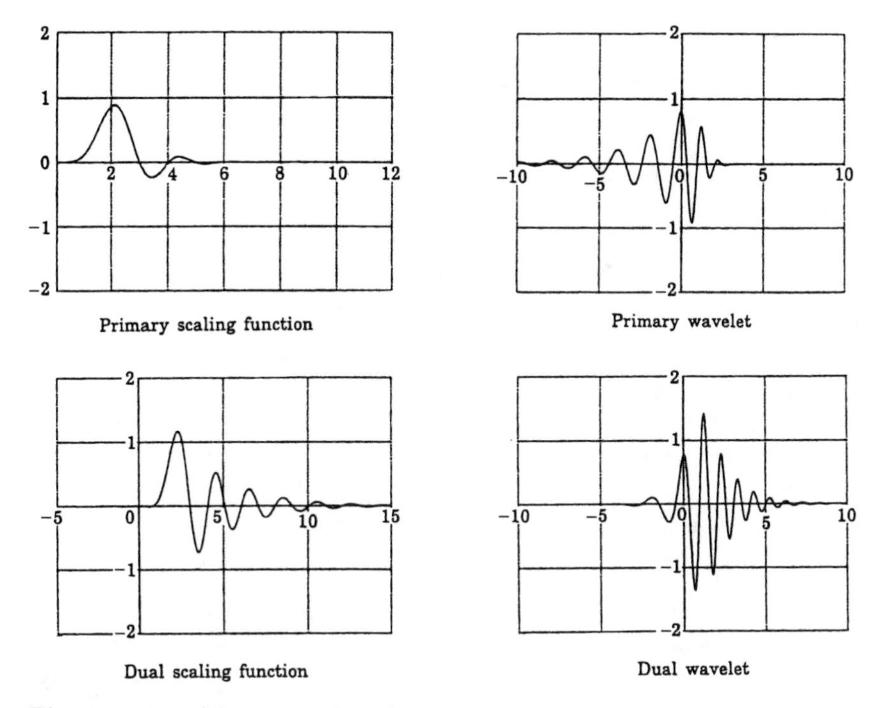

Figure 5.15. *Adaptive scaling functions and biorthogonal wavelets for the Abel transform.*

and the associate function is

$$g(\omega) = -\sum_{n=-\infty}^{\infty} |\omega + 4n\pi| \left|\hat{\phi}\left(\frac{\omega}{2} + 2n\pi\right)\right|^2.$$

Note that $g(\omega) = \overline{g(\omega)}$. This associate function and the primary symbol yield a smooth dual symbol $\tilde{m}_0(\omega)$; therefore the representation matrix of the Hilbert transform using adaptive biorthogonal wavelets can be made block diagonal. The adaptive scaling functions and the biorthogonal wavelets are shown in Figure 5.14.

The **Abel transform**, defined by

$$A[u(x)] = \int_{-\infty}^{x} \frac{u(y)}{\sqrt{x-y}}\, dy$$

[19], is invariant to scale transformations

$$A[u(\lambda\cdot)](x) = \lambda^{-1/2}\, A[u(\cdot)](\lambda x).$$

Its Fourier transform is given by

$$\hat{A}[u(\cdot)](\omega) = \sqrt{\frac{\pi}{2}}\ (1 - sgn[\omega]i)\ \frac{\hat{u}(\omega)}{\sqrt{|\omega|}},$$

and the associate function by

$$g(\omega) = \sqrt{\frac{\pi}{2}} \sum_{n=-\infty}^{\infty} \frac{1 - sgn[\omega + 4n\pi]i}{\sqrt{|\omega + 4n\pi|}}\ \left| \hat{\phi}\left(\frac{\omega}{2} + 2n\pi\right) \right|^2.$$

We use the primary symbol $m_0(\omega)$ to show that the dual $\tilde{m}_0(\omega)$ is well defined. Since the associate function $g(\omega)$ is not necessarily real or pure imaginary, the representation matrix of the Abel transform using adaptive biorthogonal wavelets is semi-block-diagonal. The adaptive scaling functions and the biorthogonal wavelets are shown in Figure 5.15.

5.4.4. Linear Combinations of Block-Diagonal Operators

In this subsection, we use the techniques developed above and show how the inverse operators of Hilbert transforms and odd-order differential operators can be decomposed into two operators which can be block diagonalized using adaptive biorthogonal wavelets. These operators appear in many scientific and engineering problems.

The **Hilbert transform** is defined by

$$f(x) = H[u(x)] = \frac{1}{\pi}\ vp \int_{-\infty}^{\infty} \frac{u(y)\ dy}{y - x}. \tag{5.19}$$

Let H_+ and H_- denote the operators

$$H_{\pm}[u(x)] = H[u(x)]\ \pm\ \beta\ iu(x), \tag{5.20}$$

where $\beta > 0$ is a real constant. These operators are invariant to scale transformations

$$H_{\pm}[u(\lambda\cdot)](x) = H_{\pm}[u(\cdot)](\lambda x),$$

and their Fourier transforms are

$$\begin{aligned}\widehat{H_+}[u(\cdot)](\omega) &= i\ J_+(\beta)\ \hat{u}\ (\omega),\\ \widehat{H_-}[u(\cdot)](\omega) &= i\ J_-(\beta)\ \hat{u}\ (\omega),\end{aligned}$$

where

$$J_+(\beta) = \begin{cases} \beta + 1 & \text{for} \quad \omega > 0,\\ \beta - 1 & \text{for} \quad \omega < 0.\end{cases}$$

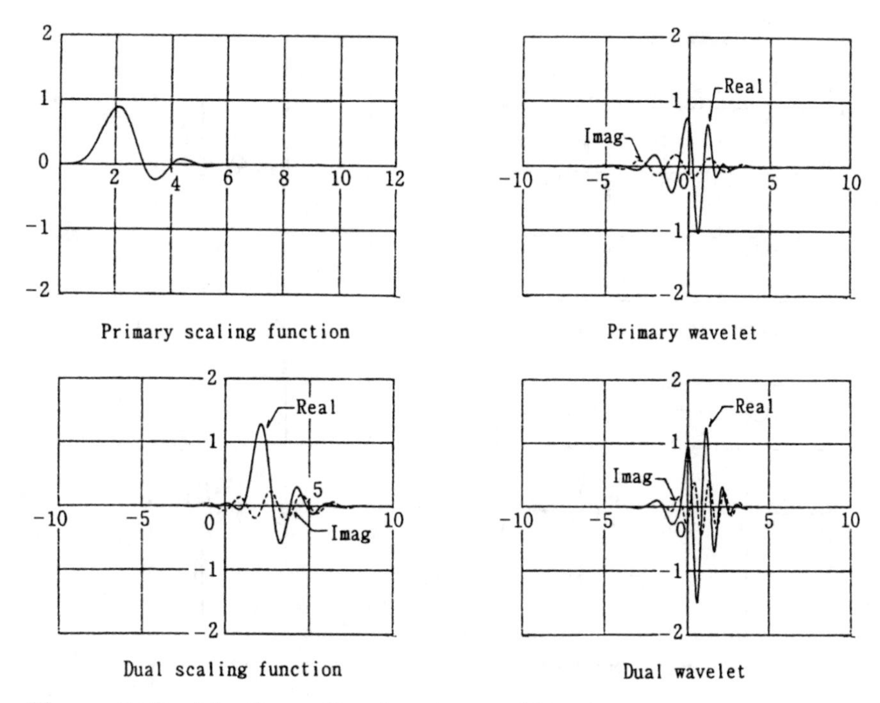

Figure 5.16. *Adaptive scaling functions and biorthogonal wavelets for* $H_{\pm}$.

$$J_{-}(\beta) = \begin{cases} 1-\beta & \text{for} \quad \omega > 0, \\ -\beta - 1 & \text{for} \quad \omega < 0. \end{cases}$$

The associate functions $g_{\pm}(\omega)$ of the operators $H_{\pm}$ are

$$g_{\pm}(\omega) = \sum_{n=-\infty}^{\infty} i \; J_{\pm}(\beta) \; \left| \hat{\phi}\left(\frac{\omega}{2} + 2n\pi\right) \right|^2 .$$

We use the primary symbol $m_0(\omega)$ to determine the dual symbol $\tilde{m}_0(\omega)$, which is well defined. The representation matrices of the operators H_+ and H_- using adaptive biorthogonal wavelets are block diagonal. We use equations (5.19) and (5.20) to rewrite $f(x)$ as

$$f(x) = H_+[u(x)] - \beta \; i \; u(x),$$

to show explicitly how the Hilbert transform can be expressed as a sum of block-diagonalizable operators. The inverse Hilbert transform is $-H$ so that

$$u(x) = -H_+[f(x)] + \beta \; i \; f(x).$$

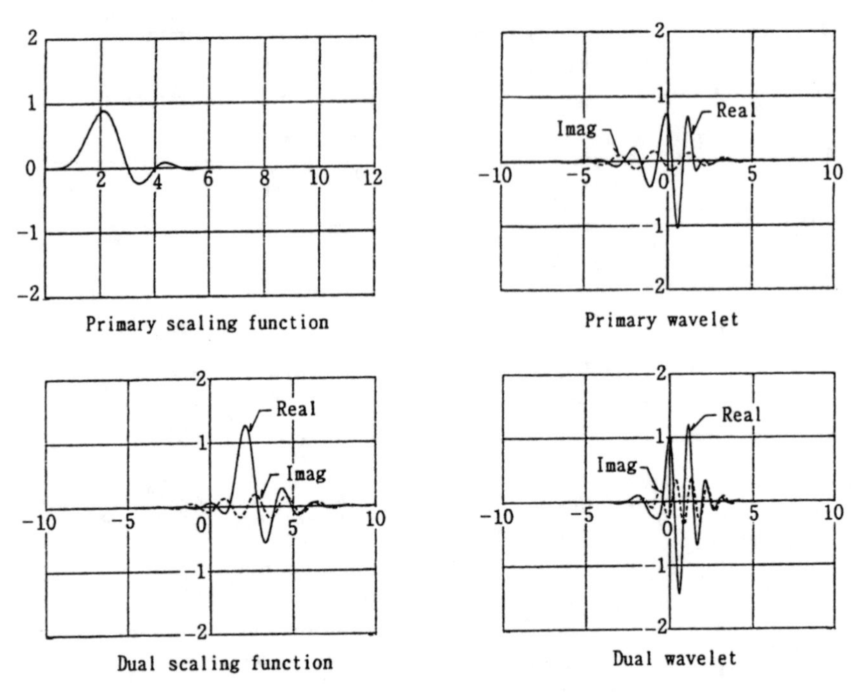

Figure 5.17. *Adaptive scaling functions and biorthogonal wavelets for* $(I_{1/2})^{2l}\ H_{+}$.

The adaptive scaling functions and biorthogonal wavelets are shown in Figure 5.16.

Consider the odd-order differential equation

$$A\ u(x) = f(x),$$

where $A = d^l/dx^l$ and l is an odd integer. The Fourier transform of the inverse of A, the Fourier transforms of the Riesz potential, and Hilbert transform are

$$\begin{aligned} \widehat{A^{-1}}\ [f(\cdot)]\ (\omega) &= \hat{f}(\omega)/(i\omega)^l, \\ \widehat{I_\alpha}\ [f(\cdot)]\ (\omega) &= \hat{f}(\omega)/|\omega|^\alpha, \\ \hat{H}\ [f(\cdot)]\ (\omega) &= i\ sgn(\omega)\ \hat{f}(\omega), \end{aligned}$$

respectively. We apply $(I_{1/2})^{2l}\ H$ to $f(x)$ to obtain

$$\left(\widehat{I_{1/2}}\right)^{2l}\ H[f(\cdot)]\ (\omega) = i\ \hat{f}\ (\omega)/\omega^l.$$

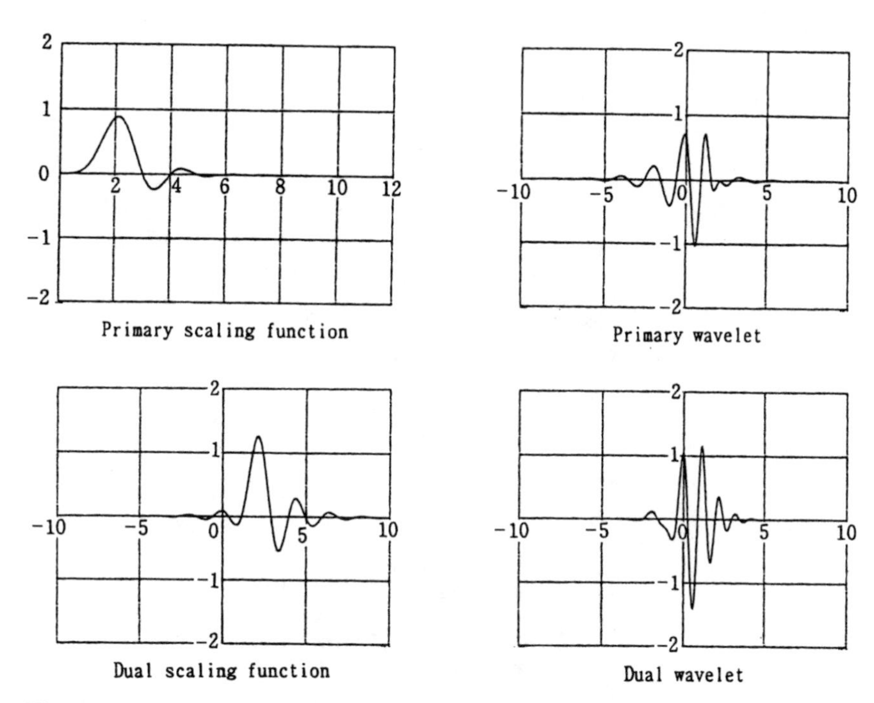

Figure 5.18. *Adaptive scaling functions and biorthogonal wavelets for* $(I_{1/2})^{2l}$.

Then the solution $u(x)$ of the inverse problem associated with the odd-order differential equation is

$$\begin{aligned} u(x) &= (1/i^{l+1})(I_{1/2})^{2l} H[f(x)] \\ &= (1/i^{l+1})\left((I_{1/2})^{2l} H_+[f(x)] - \beta i\, (I_{1/2})^{2l}[f(x)]\right). \end{aligned} \tag{5.21}$$

The operators $(I_{1/2})^{2l}\ H_+$ and $(I_{1/2})^{2l}$ satisfy the conditions from subsection 5.4.3 which guarantee a block-diagonal representation matrix of the operators using adaptive biorthogonal wavelets. The adaptive scaling functions and the biorthogonal wavelets of the operator $(I_{1/2})^{2l} H_+$ are shown in Figure 5.17, and the operator $(I_{1/2})^{2l}$, the adaptive scaling functions, and biorthogonal wavelets are shown in Figure 5.18.

To summarize, given the system $Au(x) = f(x)$ and $u(x)$, to determine $f(x)$, calculate the following:

(i) $\mathbf{c_j}$ from $u(x)$,
(ii) $\mathbf{c_{j-1}}$, $\mathbf{d_{j-1}}$, $\cdots$ from $\mathbf{c_j}$,

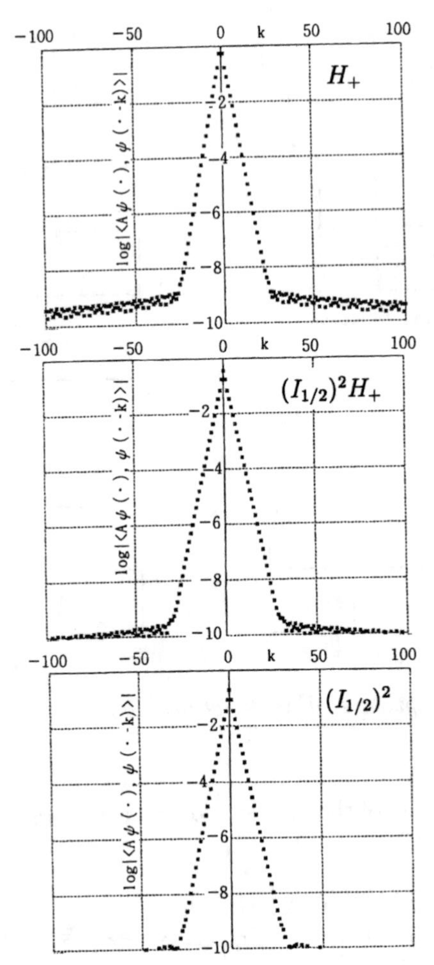

Figure 5.19. *Absolute values of the inner product* $\langle A\ \psi(\cdot),\ \psi(\cdot - k)\rangle$.

(iii) $\tilde{\mathbf{t}}_{\mathbf{j-1}}$ from $\tilde{\mathbf{t}}_{\mathbf{j-1}} = \Gamma_{\mathbf{j-1}}\ \mathbf{d}_{\mathbf{j-1}}$,
(iv) $\tilde{\mathbf{s}}_{\mathbf{j}}$ from $\tilde{\mathbf{t}}_{\mathbf{j-1}}$, and
(v) $f(x)$ from $\tilde{\mathbf{s}}_{\mathbf{j}}$.

Our procedure can also be applied to differential operators of even order; however, for even-order operators, Dahlke and Weinreich's method [2] is recommended. Williams and Amaratunga [21] note that, except for the third step, the number of operations in each step is of order $O(n)$ so the total number of operations to determine $f(x)$ depends on the computation of $\tilde{\mathbf{t}}_{\mathbf{j-1}} = \Gamma_{\mathbf{j-1}}\mathbf{d}_{\mathbf{j-1}}$. Recall from subsection 5.4.1 that the elements of the matrix $\Gamma_{\mathbf{j}}$ are $\langle A\psi(\cdot),\ \psi(\cdot - k)\rangle$. The absolute values of this inner product

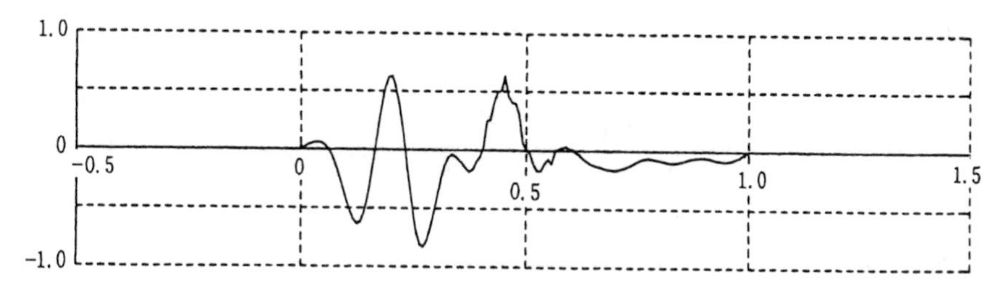

Figure 5.20. *Model data with noise.*

for the operators H_+, $(I_{1/2})^{2l}H_+$, and $(I_{1/2})^{2l}$ are shown in Figure 5.19. The values of the inner product are less than the order 10^{-6} for $|k| > 20$. In practice, solutions are within a small error when the bandwidth of the matrix Γ_j is limited to $|k| < 20$ so that the number of the operation in the third step, and the total, is $O(n)$.

5.4.5. Data Correction Using Adaptive Biorthogonal Wavelets

In this subsection, we propose a new method for correcting noisy data in the time-frequency domain using biorthogonal wavelets. Our model data (shown in Figure 5.20) are contaminated with two different types of noise. To determine a good approximation to clean data, i.e., data with smooth oscillations and a vanishing integral, we use the biorthogonal wavelets of the inverse of the operator (d^l/dx^l) described in the previous subsection.

One type of noise from the data is removed by expanding the data using biorthogonal wavelets adapted to $(I_{1/2})^2H_+$. The expansion coefficient $\alpha_{j,k}$ represents the intensity of the component at scale j and time $t = (k/2^j)$. In regions for which scientific evidence suggests the existence of contamination, we set the wavelet coefficients to be zero. For the simple example presented in this paper, we set $\alpha_{j,k} = 0$ for large j and $(k/2^j) \approx 0.5$. Physical, observational, and instrumental conditions as well as the desired level of accuracy and noise reduction must be carefully considered to determine appropriate noise reduction criteria in this step. Corrected data are computed by taking the inverse wavelet transform. Figure 5.21 shows how the noise near the center almost disappears, and the data change very little at other times.

Let $x(t)$ denote the data after this initial noise removal step. If the integral of $x(t)$ does not vanish, i.e.,

$$\int_{-\infty}^{\infty} x(t)\, dt \neq 0,$$

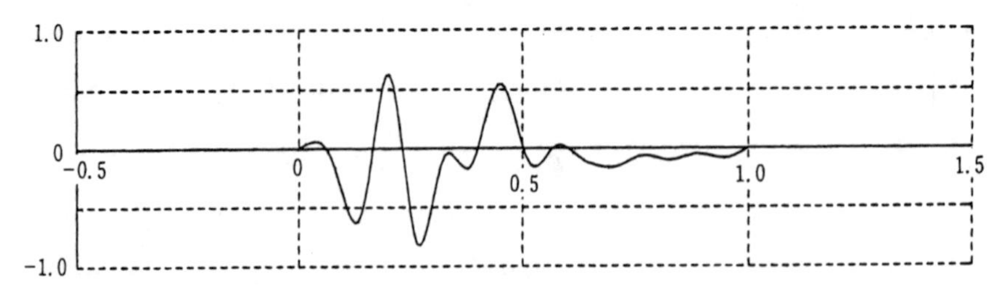

Figure 5.21. *Model data after noise removal.*

we must determine further corrected data $x_{ap}(t)$ so that it does vanish. x_{ap} must satisfy

$$\int_{-\infty}^{\infty} x_{ap}(t)\, dt = 0; \qquad \text{the DC component} = 0, \tag{5.22}$$

$$x_{ap}(t) = 0; \qquad t < 0 \quad \text{or} \quad t > 1, \tag{5.23}$$

since the velocity before and after an earthquake must be zero. Some simple noise reduction techniques do not yield an appropriate solution to our problem. Straightforward subtraction of the average of $x(t)$ over the interval $[0, 1]$ from $x(t)$ will not appropriately yield $x_{ap}(t)$, because discontinuities in the data will appear at $t = 0$ and $t = 1$—neither will the addition of a polynomial to the data, because regularity properties will not be preserved. The correction method we propose removes the DC component, preserves regularity properties, and sets the signal to zero outside of the interval under consideration.

Let F be the linear operator

$$F\ \psi_{j,k} = \frac{1}{B_{j,k}}\ \tilde{\psi}_{j,k},$$

where $B_{j,k}$ is a positive constant which depends on j and k. We define a new norm $\langle \cdot\,, \cdot \rangle_N$ such that

$$\langle\, f,\, g\, \rangle_N = \int_{-\infty}^{\infty} f(t)\ \overline{(Fg(t))}\ dt.$$

$x_{ap}(t)$ can be expanded in terms of biorthogonal wavelets as

$$x_{ap}(t) = \sum_{j>0} \sum_{k} \alpha_{j,k}\ \psi_{j,k}(t). \tag{5.24}$$

We assume that the data are periodic with period $[t_0, t_1]$, which contains the unit interval. Since the wavelets do not have a DC component, (5.22) is satisfied. The biorthogonal wavelet transform of $x(t)$ in the interval $[t_0, t_1]$ is

$$x(t) = \sum_{j>0} \sum_{k} \beta_{j,k}\ \psi_{j,k}(t) + \gamma_{0,0}\ \phi(t).$$

We rewrite condition (5.23) as

$$x_{ap}(t_l) = \sum_{j>0} \sum_{k} \alpha_{j,k}\ \psi_{j,k}(t_l) = 0 \quad \text{for} \quad t_l \notin [0,1], \tag{5.25}$$

and choose $x_{ap}(t)$ to minimize the norm

$$\langle\, x_{ap}(t) - x(t),\ x_{ap}(t) - x(t)\,\rangle_N.$$

$x_{ap}(t)$ is determined using Lagrange's method of indeterminate coefficients, in which the error function is defined as

$$\Phi(\alpha_{j,k}) = \langle\, x_{ap}(t) - x(t),\ x_{ap}(t) - x(t)\,\rangle_N - \sum_{l} \lambda_l \sum_{j>0} \sum_{k} \alpha_{j,k}\psi_{j,k}(t_l),$$

and λ_l is the indeterminate coefficient. To determine wavelet coefficients and indeterminate coefficients which minimize the error function Φ, we set

$$\frac{\partial \Phi}{\partial \alpha_{j,k}} = \frac{\partial \Phi}{\partial \overline{\alpha}_{j,k}} = 0$$

and sort terms to obtain

$$\alpha_{j,k} = \beta_{j,k} + B_{j,k} \sum_{l} \lambda_l\ \overline{\psi}_{j,k}(t_l). \tag{5.26}$$

Multiplication by $\psi_{j,k}(t_l)$ leads to

$$\sum_{j,k} \beta_{j,k}\ \psi_{j,k}(t_l) + \sum_{j,k} \psi_{j,k}(t_l)\ B_{j,k} \sum_{l'} \lambda_{l'}\ \overline{\psi}_{j,k}(t_{l'}) = 0,$$

which yields the indeterminate coefficient λ_l. The wavelet coefficient $\alpha_{j,k}$ is then determined from (5.26) and the corrected data $x_{ap}(t)$ from (5.24).

The constant $B_{j,k}$ represents the size of the correction at scale j and time $t = k/2^j$. For a large correction in $\alpha_{j,k}$ with large amplitude, set $B_{j,k} = |\beta_{j,k}|$ to indicate that portions of the data with a large amplitude in the time-frequency domain must be corrected to a relatively large degree. For small corrections in portions with large amplitude, set $B_{j,k} = 1/|\beta_{j,k}|$. For uniform corrections, set $B_{j,k} = 1$.

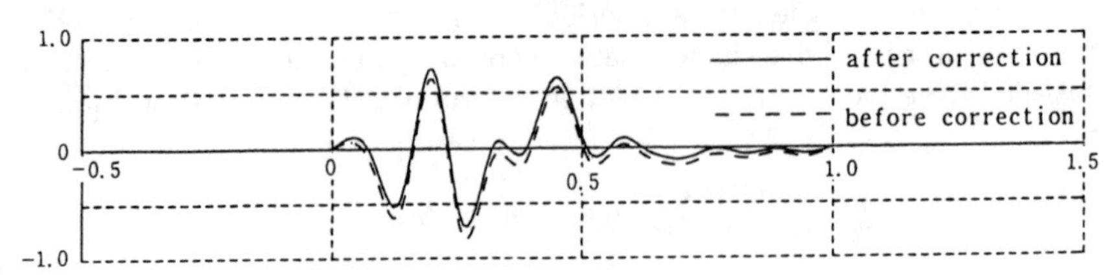

Figure 5.22. *Data x_{ap} after base line correction.*

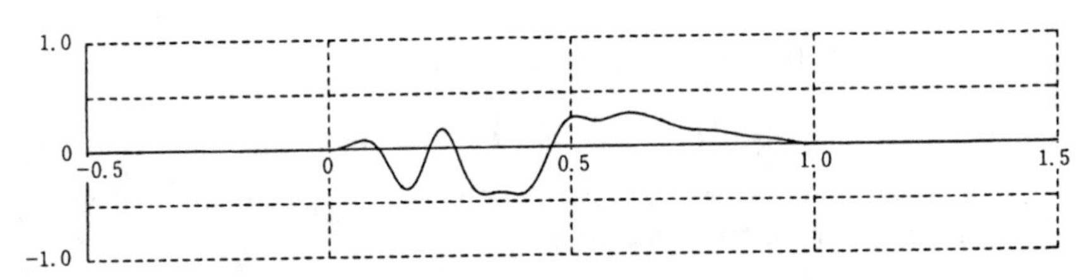

Figure 5.23. *Data obtained by integrating x_{ap}.*

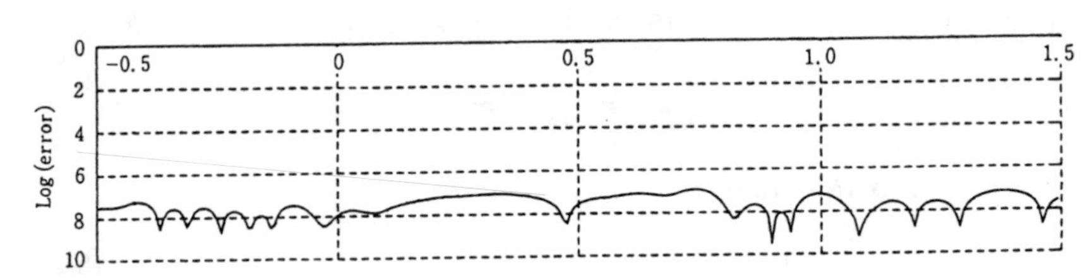

Figure 5.24. *Log-error in x_{ap}.*

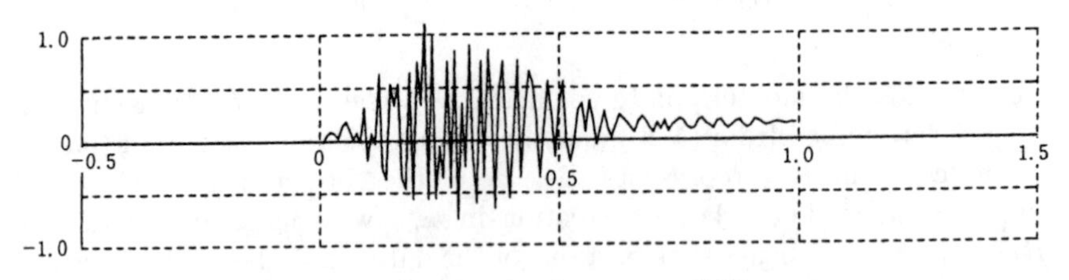

Figure 5.25. *Original wave + shift wave.*

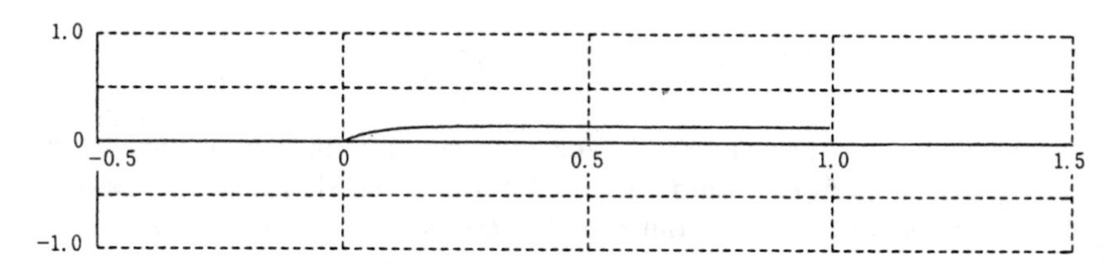

Figure 5.26. *Shift wave.*

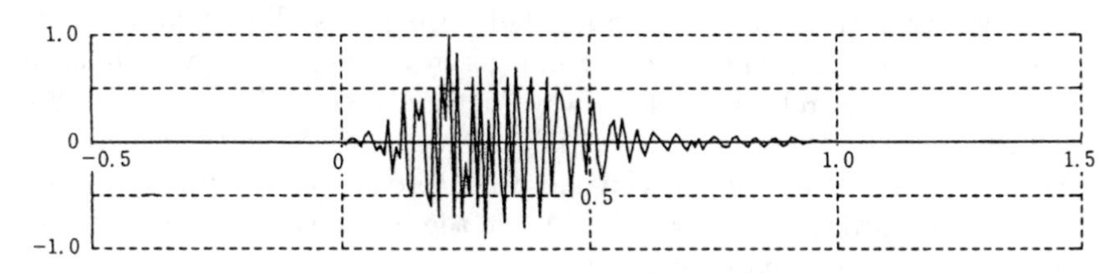

Figure 5.27. *Original wave.*

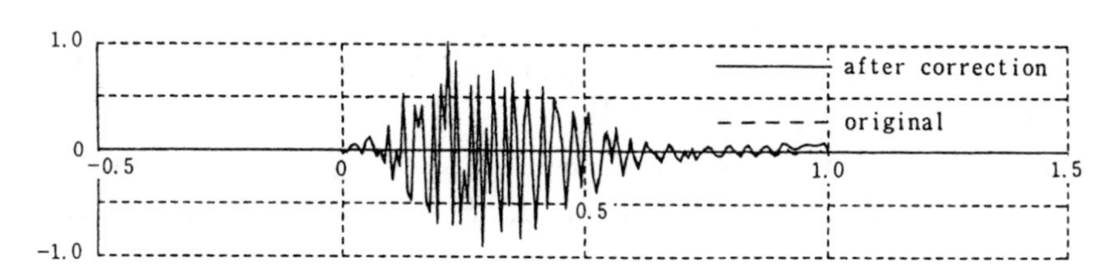

Figure 5.28. *Comparison of x_{ap} and the original wave.*

Figure 5.21 illustrates the data $x(t)$ after the first noise removal step, with DC component

$$\int_{-\infty}^{\infty} x(t)\,dt \;\cong\; 0.7.$$

We set $B_{j,k} = |\beta_{j,k}|$ and apply the second correction method in the interval $[t_0, t_1] = [-0.5, 1.5]$ to obtain corrected data $x_{ap}(t)$, shown in Figure 5.22. The value of $x_{ap}(t)$ is zero outside the interval $[0, 1]$, and the magnitude of the correction is large when there are large oscillations. Results from computing the integral of $x_{ap}(t)$ from (5.21) using only the matrix elements of $\Gamma_{\mathbf{j}}$, where $|k| < 20$, are shown in Figure 5.23. To determine the accuracy of the integral, we plot the error from the theoretical solution which is of order 10^{-7} (see Figure 5.24). Since the magnitude of the solution is of order 10^{-1}, the relative error is 10^{-6}, so the accuracy of our solution agrees with that of the matrix $\Gamma_{\mathbf{j}}$.

In our final experiment, we apply our correction method to acceleration data, which are artificially contaminated with a base line shift in time, shown in Figure 5.25. The wave is produced by adding a shift wave (shown in Figure 5.26) to relatively clean wave data (shown in Figure 5.27). We set $[t_0, t_1] = [-0.5, 1.5]$ and $B_{j,k} = |\beta_{j,k}|$. Figure 5.28 shows the wave after correction (solid line) and the original wave (broken line). Although there is a slight upward tilt from $t = 0.9$ to 1.0 in the corrected wave, the original and corrected waves match very well.

Acknowledgements. The authors would like to express their sincere thanks to Dr. Mei Kobayashi for her critical reading of the manuscript and helpful comments.

References

[1] K. Amaratunga et al. (1994), "Wavelet-Galerkin solutions for one dimensional partial differential equations," *Internat. J. Numer. Methods Engrg.*, vol. 37, pp. 2703–2716.

[2] S. Dahlke, I. Weinreich (1993), "Wavelet-Galerkin methods: An adapted biorthogonal wavelet basis," *Constr. Approx.*, vol. 9, pp. 237–262.

[3] W. Dahmen, A. Kurdila, P. Oswald (eds.) (1997), *Multiscale Wavelet Methods for Partial Differential Equations*, Academic Press, Tokyo.

[4] I. Daubechies (1992), *Ten Lectures on Wavelets*, SIAM, Philadelphia, PA.

[5] R. Glowinski, W. Lawton, M. Ravachol, E. Tenenbaum (1989), "Wavelet solution of linear and nonlinear elliptic, parabolic and hyperbolic problems in one space dimension," *Aware Inc. Technical Report* AD890527.1, Boston, MA, pp. 1–79.

[6] M. Higuchi, M. Yamada, Y. Mitsuta (1991), "Local high wind in Amarube Valley," *Proc. Disaster Prevention Research Inst., Kyoto Univ.*, vol. 34, B–1, pp. 13–18 (in Japanese).

[7] L. Jameson (1993), "On the wavelet-based differentiation matrix," *J. Sci. Comput.*, vol. 8, pp. 267–305.

[8] L. Jameson (1994), "On the wavelet-optimized finite difference method," *ICASE Report* 94-9, Hampton, VA.

[9] L. Jameson (1995), "On the spline-based wavelet differentiation matrix," *Appl. Numer. Math.*, vol. 17, pp. 33–45.

[10] L. Jameson (1996), "The differentiation matrix for Daubechies-based wavelets on an interval," *SIAM J. Sci. Statist. Comput.*, vol. 17, pp. 498–516.

[11] P. Kailasnath, K. Sreenivasan, G. Stlovitzky (1992), "Probability density of velocity increments in turbulent flows," *Phys. Rev. Lett.*, vol. 68, pp. 2766–2769.

[12] A. Latto, E. Tenenbaum (1990), "Compactly supported wavelets and the numerical solution of Burger's equation," *C. R. Acad. Sci. Paris*, Ser. I, vol. 311, pp. 903–909.

[13] Y. Maday, V. Perrier, J. Ravel (1991), "Dynamical adaptivity using wavelets basis for the approximation of partial differential equations," *C. R. Acad. Sci. Paris*, Ser. I, vol. 312, pp. 405–410.

[14] S. Mallat (1989), "A theory for multiresolution signal decomposition: The wavelet representation," *IEEE Trans. Pattern Anal. Mach. Intell.*, vol. 11, pp. 674–693.

[15] Y. Meyer (1989), "Orthonormal wavelets," pp. 21–37 in *Wavelets*, J. Combes, A. Grossmann, Ph. Tchamitchian (eds.), Springer-Verlag, Tokyo.

[16] S. Qian, J. Weiss (1993), "Wavelets and the numerical solution of partial differential equations," *J. Comp. Physics*, vol. 106, pp. 155–175.

[17] F. Sasaki, M. Yamada (1997), "Biorthogonal wavelets adapted to integral operators and its applications," *Japan J. Indust. Appl. Math.*, vol. 14, pp. 257–277.

[18] E. Stein (1970), *Singular Integrals and Differentiability Properties of Functions*, Princeton Univ. Press, Princeton, NJ.

[19] F. Tricomi (1957), *Integral Equations*, Wiley Interscience, New York.

[20] J. Weiss (1991), "Wavelets and the study of two dimensional turbulence," *Aware Inc. Technical Report AD*910628, Boston, MA, pp. 1–28.

[21] J. Williams, K. Amaratunga (1995), "A multiscale wavelet solver with $O(n)$ complexity," *J. Comp. Phys.*, vol. 122, pp. 30–38.

[22] M. Yamada, M. Higuchi, T. Hayasi, Y. Mitsuta (1990), "Wavelet analysis of high wind," *Proc. Disaster Prevention Research Inst., Kyoto Univ.*, vol. 33, B-1, pp. 285–295 (in Japanese).

[23] M. Yamada, K. Ohkitani (1990), "Orthonormal wavelet expansion and its application to turbulence," *Progr. Theoret. Phys.*, vol. 83-5, pp. 819–823.

[24] M. Yamada, K. Ohkitani (1991), "Orthonormal wavelet analysis of turbulence," *Fluid Dynam. Res.*, vol. 8, pp. 101–115.

[25] M. Yamada, K. Ohkitani (1991), "An identification of energy cascade in turbulence by orthonormal wavelet analysis," *Progr. Theoret. Phys.*, vol. 86-4, pp. 799–815.

[26] M. Yamada, S. Kida, K. Ohkitani (1993), "Wavelet analysis of PDFs in turbulence," pp. 188–199 in *Unstable and Turbulent Motion of Fluids*, S. Kida (ed.), World Scientific, Singapore.

Index